Hugh Barker is a non-fiction author and editor; as the latter he has edited several successful popular maths books, including *A Slice of Pi*. He is the author of *Million Dollar Maths* (Atlantic Books, October 2018) and *High Tech Maths* (Atlantic, 2019–20). Hugh is a keen amateur mathematician, and was accepted to study maths at Cambridge University aged sixteen. Japanese rights for *Million Dollar Maths* recently sold for £21,000 after a three-way auction.

Praise for *Million Dollar Maths*

'Great fun. A clear, original and highly readable account of the curious relationship between mathematics and money.'
Professor Ian Stewart – author of *Significant Figures*

'A lively crash course in the mathematics of gambling, investing, and managing. Hugh Barker makes deep ideas fun and profitable.'
William Poundstone – author of *How to Predict the Unpredictable*

Also by Hugh Barker

Million Dollar Maths
The Forking Trolley

Lying Numbers

How Maths and Statistics are Twisted and Abused

HUGH BARKER

ROBINSON

ROBINSON

First published in Great Britain
in 2020 by Robinson

3 5 7 9 10 8 6 4

Copyright © Hugh Barker, 2020

A CIP catalogue record for this book
is available from the British Library.

ISBN: 978-1-47214-361-7

Typeset in Scala by Hewer Text UK Ltd
Printed and bound in Great Britain
by Clays Ltd, Elcograf S.p.A.

Papers used by Robinson
are from well-managed forests
and other responsible sources.

Robinson
An imprint of
Little, Brown Book Group
Carmelite House
50 Victoria Embankment
London EC4Y 0DZ

An Hachette UK Company
www.hachette.co.uk

www.littlebrown.co.uk

I'd like to dedicate this book to Diane and Leah, for helpful suggestions, feedback, endless patience and cups of coffee, among much, much more.

Contents

Introduction

THE NUMBERS DON'T LIE

Most people are liars. It's a sad fact, but statistically it has been proven that 97 per cent of men and 96 per cent of women tell at least three lies a day. Furthermore, in any single year, adults tell an average of 589 lies, ranging from white lies to partial truths to raging whoppers.

OK, I was lying there, I made up those facts. They didn't even make any mathematical sense. They weren't based on any actual research: it is what the experts refer to as 'bullshit'. I promise if I tell any more lies in this book, I will own up to them. It's just an example of the most basic way that numbers can be used deceptively, which is by simply making the figures up as you go along.

However, you can use genuine figures and research and still find ways to mislead people – you can cherry-pick useful statistics, design misleading visual imagery (such as poorly constructed graphs), compare apples with oranges, give only part of the truth, or exploit people's natural cognitive biases and fallacies. The numbers don't lie, but people do: there is a wide variety of ways that people using numbers can either lie about what the numbers say or use the data in a creative way to mislead an audience.

Here's a real bit of research from a 2018 MIT study into fake news on social media (the study was specifically based on Twitter). The researchers found that far more false stories than real ones go viral, and that on average, a false story reaches 1,500 people six times quicker than a true story does. And fascinatingly, this can't be blamed on Twitter bots, as they amplify false and true stories at the same rate – it is humans that fall for the false stories at a higher rate and are far more likely to retweet them – fake news is 70 per cent more likely to be passed on by humans.

An example is the Donald Trump tweet in 2015 claiming that 81 per cent of whites were killed by blacks (accompanied by an image citing the

Crime Statistics Bureau in San Francisco, which also claims that only 2 per cent of blacks were killed by whites). It was retweeted many thousands of times, in spite of the fact that the supposed source of the statistic doesn't exist and genuine statistics show that the percentages are completely different.

In a similar vein, InfoWars founder Alex Jones' story that a Category Six hurricane was imminent was shared two million times, even though there is no such category. Again, this was completely untrue.

Of course, entire careers have been based on lying or misrepresenting the truth. For some examples, take your pick from your favourite politicians, lawyers, spin doctors, writers, advertisers, salespeople, marketers, con artists and bankers – all of these professions will sometimes involve twisting the facts for ulterior motives. Then there are realms like economics, opinion polling and science, where it can be an essential art of self-preservation to come up with the 'right results' or to make exciting predictions that will garner attention.

The fundamental purpose of this book is to demonstrate some of the most common and devious ways in which numbers are abused and distorted. It may be that you want to learn how to use numbers dishonestly yourself or, I hope, that you are looking for a crash course in how other people might be lying to you. Alternatively, you might be interested, as I am, in the psychology of lying. Why do people at companies that commit serious frauds stay silent when they could blow the whistle? What motivates politicians when they repeat 'fake news' that they must know is untrue? Is lying a natural human instinct, and to what degree does the instinct vary among individuals? Do we all, to some extent, suffer from a tendency to self-deception and rationalisation? These are all fascinating questions that will come up along the way.

I've kept the sections fairly short on the basis that a brief, memorable example is often the best way to remember that, for instance, almost everyone has more than the average number of eyes, and that any mathematical model will output garbage if you input garbage in the first place.

If you come out of this book remembering to treat all statistics critically, and to examine the motives and methods being used by whoever is presenting you with a set of figures, then you will have learned or relearned a useful skill that can stand you in good stead when it comes to dealing with charlatans, fraudsters, con men and other ordinary members of the human race.

Failing that, it might at least remind you to never believe a word that comes out of a politician's mouth again . . .

CORONAVIRUS UPDATE

I completed work on this book a couple of months before the coronavirus outbreak of 2020, but am working on the final edit of it in April 2020, at which point the UK is in lockdown. There are a few points throughout the book where it feels like an omission, not to mention some of the bizarre and twisted ways in which data has been used both in the lead-up to the pandemic and during it. Rather than attempt to rewrite sections wholesale to reflect this, I have added a few updates throughout: I hope that by the time you are reading this the situation is under control and that the lessons are being learned in the aftermath.

CHAPTER 1

Lying Liars (and Politicians)

....

REMEMBER THAT 73.6 PER CENT OF STATISTICS ARE MADE UP

The easiest way to lie with numbers is simply to make them up. Politicians do it all the time, especially when they can't think of anything else to say. In this post-truth era it has even become a badge of honour to lie so blatantly that no one could possibly miss it. In many countries politics has become so divisive that a politician will be cheered on by their 'own side' regardless of whether they are being honest or not. It won't matter how much the 'other side' pull them up on their obvious lies, their supporters will only spur them on to keep it up.

Of course, some political lies and distortions are easier to spot than others. Since the truth-telling reputation of politicians is at an all-time low, it makes sense to start by taking a look at some of the ways that they misuse and abuse numbers and statistics.

HITTING THE HEADLINES

The entrepreneur Mark Suster tells the story of receiving a set of projections for the size of mobile-phone markets in various countries around the world. Each set of predictions showed smooth growth over successive years, but the projections were so wildly different from one another that he started to wonder how the research companies had come up with the data, and followed up with each to ask them their methodology.

The first data analyst he spoke to turned out to be a twenty-four-year-old who blithely confessed that he had been on a tight deadline and that 'my boss told me to look at the growth rate average over the past three years and increase it by 2 per cent because mobile penetration is increasing'.

Suster got similar replies from other consultants, and also confirmation that they had all projected steady growth over the period because 'nobody buys reports that just show that next year the same thing is going to happen that happened last year'.

And of course when the researchers provide their results to journalists, the effect is only amplified – the best news stories are the ones that make the boldest, most exciting claim – 'Mobile Phone Market To Double In Size Over Five Years' is a sexier story than 'Mobile Phone Market To Stay About The Same Next Year'.

So before you even start to wonder about how a statistic is being used or presented, you need to ask who produced the data, what their motives were, how large a sample they were working from, and how reliable their methods were. Statistics tend to take on a life of their own, even more so now that we have the echo chamber of social media, and one piece of melodramatic guesswork from a hungover twenty-four-year-old consultant might be next week's tabloid headline. Possibly the same kind of tabloid that will lead on 'Red Wine 80% More Dangerous Than Cocaine' one week, only to turn round two weeks later and claim that 'Red Wine Can Help Beat Dementia'.

Completely made-up stories can be even more alluring to news networks. In November 2013 it was widely reported on TV news that Samsung had paid a $1 billion fine to Apple using twenty billion nickels (five-cent coins). If the news networks had taken a few moments to do the maths and see that this would have required the entire stock of nickels in circulation, they might have bothered to check the source, which was a deliberate spoof that had been published on the internet.

> **Bullet Point: Not all statistics are made up, but the ones that are often make the best headlines.**

POTEMKIN NUMBERS

Whether you learned the art of misinformation in the debating chamber at Eton, at an advertising agency or from watching your con-man uncle trying to swindle old ladies out of their savings, you may at some point have learned how powerful a completely arbitrary number can be.

In his excellent book *Proofiness,* Charles Seife calls bogus numbers of this sort **Potemkin numbers** – after the Russian nobleman Grigory Potemkin, who had fake villages built on the banks of the Dnieper River so that he could fool Empress Catherine II into believing that his reconstruction work in the area was more advanced than it actually was. Overnight the two-dimensional constructions were packed up and moved downstream so they could be reused to fool her again.*

He gives the example of Senator Joe McCarthy, who wanted to get the attention of Congress and the media for his hysterical anti-communist crusade. So he stood up and waved a pile of paper in the air and claimed that it contained the names of 205 known communists. A week later he used the same trick and claimed there were 207 names, and the day after that he wrote to President Truman claiming he had the names of 57 commies. None of the claims had any veracity whatsoever; McCarthy simply wanted to intimidate his enemies and gain support for his witch-hunt, in which aim he was pretty successful, at least for a while.

One story that circulated widely on the internet in the years after the 9/11 attack on the Twin Towers claimed that Daisy, a golden retriever whose owner was working in the building on that day, saved her blind owner before returning several times to rescue further groups of people. The claim was that she saved about 300 people on the first run, before dashing in to save another 392 lives, and then a final trip to save 273 lives before miraculously escaping from the collapsing building suffering from exhaustion and smoke inhalation.

It's a touching story, but it's also completely untrue. There were two men who were indeed helped to safety by their guide dogs that day – Michael Hingson made it from the 78th floor with his dog Roselle and Omar Rivera from the 71st floor with his dog Salty, but neither of these two dogs ran back to save any more people.

One of the first clues that the sceptical reader should spot in the story is the suspiciously precise numbers given. It is fairly obvious that among the chaos of that day, there wasn't anyone on the ground with a clipboard taking a headcount of 273 or 392 people following a heroic dog to safety.

---

* As this is a book about lying, I had best acknowledge that this is a disputed historical anecdote, so it is best to treat it as a wonderful metaphor rather than gospel truth.

These are just Potemkin numbers that have been used to give the story an additional air of veracity and to boost its sentimental power.

Bullet Point: If you read about a heroic guide dog with super-powers, take a moment to check if the figures sound made up or not.

MASQUERADE NUMBERS

Numbers pulled out of thin air aren't always enough to persuade your audience. If you want to make a claim with very slightly more believability you might want to use a **Masquerade number**. These were originally named by the famous mathematician Carl Friedrich Gauss, when his new landlord told him his rent would be 20 florins a week but failed to mention that that was the baseline figure to which he added additional rates of 5 florins a week for maintenance, 5 florins a week for management and 5 florins a week for 'contingencies'. Making a total of 35 florins a week . . .

Actually, I'm lying again. I've no idea what rent Gauss might or might not have paid, or what currency it would have been paid in. I just made up an example that is loosely based on the way my annual service charges seem to be calculated by the freeholder. And 'Masquerade number' is a term I just invented to describe a number that is real but is being disguised as something completely different, like when someone attends a masked ball wearing a mask and disguise.

The best recent example is a claim that was made during the Brexit referendum in 2016. Vote Leave buses were emblazoned with the slogan: 'We send the EU £350 million a week. Let's fund our NHS instead.' The British are very fond of the NHS but recognise how badly it is underfunded, so this was a powerful claim, one that Vote Leave campaign director Dominic Cummings credited with swinging the final vote to a 52 per cent majority for 'Leave'. And it was repeated regularly by Leave campaigners, including Boris Johnson, both before and after the referendum.

So is there any truth in it?

Well, there is a small grain of truth in there. It is true that our notional contribution to the EU was £17 billion a year, which does come to about £350 million a week. That's the real figure behind the mask.

However, we never actually sent that much to the EU. This is because, firstly, we were subject to a negotiated rebate of £4 billion, making a total contribution of £13 billion a year, or about £250 million a week. So to start with, imagine you have a supermarket bill for a shopping trip in which every item you bought was on a three-for-two offer. If the original total on the bill was £150 you might actually be charged £100 after the discounts were applied. The claim on the bus was equivalent to saying that you actually gave the supermarket £150, when you clearly didn't.

In addition, there was about £4 billion of subsidies to poor areas, agriculture and so on, making the total net contribution closer to £9 billion a year, or about £175 million a week.

Now, to acknowledge that the politics of this situation are more complicated, there are all kinds of additional arguments that could be made at the time of writing (when the situation is still unresolved three years after the referendum) on either side of the debate. Brexiteers might claim that the advantages of leaving mean we will gain more financially than this suggests. Remainers would argue that the economic benefits we receive in exchange for our contribution will make the cost of leaving higher. And they might also point out that the original claim on the bus includes an additional unjustified assumption, which is that every penny of money that we didn't send to the EU would be spent on the NHS.

And so on and so on until we all wish we had never even heard the word 'Brexit' in the first place.

But none of this affects the core claim, which was a simple statement about the amount we actually sent at the time of the referendum to the EU on a weekly basis. And in that respect, what we were looking at was a real figure that was heavily disguised to make it far more persuasive than it actually was. And this is significant because, while the UK Statistics Authority has repeatedly described the claim as 'misleading', it has continued to be cited, and 2018 research showed that two years after the original vote, 42 per cent of people who had heard the original claim thought it was true, 36 per cent thought it was false, and the remaining 22 per cent were unsure.

So nearly two-thirds of people who had heard the claim thought it was definitely or possibly true, in spite of it being shown repeatedly to be untrue.

That is the danger of a big, simple lie like that.

Bullet Point: As the saying goes, a lie can be halfway around the world before the truth has even got out of bed.

Do We Even Want to Know the Truth?

This might seem an odd question, but one of the first questions you should ask yourself is whether what you want out of a particular statistic or set of statistics is the truth or validation. Bear in mind that we live in a world of fake news and echo chambers, and in many cases the source you choose for your news will define the kind of news you are going to hear.

Dan Kahan, a professor who studies cognition at Yale, has discussed the subject of **motivated numeracy**: he set up an experiment that started out by giving all the participants a maths test, to establish their basic level of numeracy, along with a quiz to identify their basic political leanings. Then he gave them a logical problem in which they looked at some data and had to identify whether or not a fictional skin cream was effective or not.

The question was a moderately difficult one but one that could be said to have a correct answer, and the results were much as one would expect. The subjects with the highest levels of numeracy were more likely to get the results right than those with lower levels of numeracy.

Then he gave them a second problem that was cunningly framed in exactly the same way, but this time the question was about whether a law that banned people from carrying a concealed gun in public would make crime go up or down. This time the results were sharply different. Rather than the replies being more likely to be mathematically correct among the group with higher numeracy, the results were fairly strongly divided along political lines. The group who were predisposed to believe gun control was a good thing thought crime would go down, whereas the group who were predisposed to believe the opposite gave the opposite interpretation of the data.

The really interesting thing is that the levels of numeracy made little to no difference to how likely the participants were to get the answer right. The participants with the highest levels of numeracy were just as likely to get the answer wrong as those with low levels of numeracy.

The experiment has been replicated by academics in Australia, where rather than using gun control as the trigger, they used the question of whether or not closing down a local coal-fired power station would reduce carbon emissions in the region. In this case the Green Party supporters were more likely to predict it would than the One Nation supporters (bearing in mind that this is a subject that is especially controversial in Australia as coal mining is being increased at the same time as the Great Barrier Reef is being directly harmed by carbon emissions, and global warming is blamed by many for the increase in bushfires). And again, the level of numeracy of the subject made little to no difference when it came to getting the answer right.

I tried a small experiment of my own for fun: I have a large American family that includes both Trump supporters and fierce critics of the President. I wondered if warning people of the purpose of the test would make any difference, so I pointed out in advance that I was writing a book about how statistics were abused and I was wondering whether or not the answers to a problem would vary with political bias. I gave an intentionally fictional set of statistics in which I gave GDP (gross domestic product) figures for the US in 2017, 2018 and 2019 as being $100 trillion, $101 trillion and $102 trillion. Then I asked them to predict the GDP for 2020.

Now, there is no 'right' answer to this question. Simple projection of a trend would suggest that $103 trillion might be a sensible guess, but given how complicated economics is, there really isn't enough data here to make a prediction. However, I took $103 trillion as being the 'neutral' answer to the question.

Sure enough, every Trump supporter gave an answer that was higher than $103 trillion, while every critic gave an answer that was lower. The one exception was a cousin who used to be a Republican, but has become entirely disenchanted with Trump and is now effectively agnostic. Perhaps that was why he simply looked at the actual numbers and saw that the logical guess was simply to project the same expansion for 2020.

Now, this was only a small group, and I didn't give them any maths tests first, but it is another reminder of how significant our preconceived opinions are when it comes to looking at the numbers. Basically we find it very hard to be dispassionate or neutral when it comes to statistics about something we care deeply about.

I tend to explain this as being a result of **cognitive dissonance.** This is the psychological quirk that we have whereby we find it uncomfortable to believe two logically incompatible things at the same time. When we are confronted with such discomfort, we tend to jettison whichever of the two beliefs we have the least attachment to. When it comes to a neutral question like the one about skin cream, we can apply our common sense and see the correct answer. But when it comes to data that conflict with our core political beliefs we choose to jettison that common sense and resort to answers that validate our expectations. It's hard to overcome this psychological tendency, but the very least we can do is own up to the fact we aren't always as rational as we think we are, especially when it comes to subjects on which we have strong beliefs.

On the bright side, this is part of what makes us human: we have strong opinions, emotions and faiths. On the down side, that means we can sometimes behave remarkably irrationally when we are trying to understand the numbers.

Bullet Point: As the Dean Friedman song says, 'We can thank our lucky stars that we're not as smart as we like to think we are.'

The Fine Art of Disestimation

Another useful term introduced by Charles Seife is **disestimation**, which is essentially the art of leaping to conclusions from a set of statistics that at best give us a fuzzy picture of reality. He has used this term in reference to a 2010 survey by the Pew Forum on Religion and Public Life. The survey led to headlines such as 'Atheists Know More About Religion

Than Believers', based on the fact that atheists and agnostics had accurately (on average) answered more questions than those of religious faith.

I'll discuss Seife's analysis of the shortcomings of this conclusion in a moment. But first, I think it is worth examining an immediate riposte that came in a 2010 article by Bret Feiler on the Fox News website. Now, I don't think it is controversial to suggest that Fox News generally tends to defend conservative, traditional and religious view-points, so I'm going on the assumption that their motivation was to debunk the survey.

The article is entitled 'We Didn't Flunk the Religion Test – 4 Important Truths About Americans and God', and it essentially goes into the numbers in an attempt to suggest that the headlines crediting atheists with more religious knowledge than believers were flawed.

First the author pointed out that the general level of knowledge among the group polled was fairly low since as few as 59 per cent knew who the vice president at the time was or could correctly identify what an antibiotic is. By contrast 63 per cent could name the first book of the Bible, and 68 per cent knew that the American constitution forbade the establishment of religion. And while only that 59 per cent could name Joe Biden, the same percentage could name the Koran as being the holy book of Islam.

His rather questionable conclusion from these percentages (which aren't, after all, that different to one another) was that the group surveyed weren't very knowledgeable in general but that, by contrast, they knew more about religion: indeed, according to the author, this means that 'Americans are religious savants'.

Let's call that the '**Point at something else and shout about it' argument**.

Feiler's second argument was that 'the most popular religious figure in America is Moses'. He does give statistical support for that statement, as more people correctly answered questions about the stories of Moses and the Ten Commandments than about other biblical figures. (I'm going to go ahead and blame Charlton Heston for that quirky fact, although it is also possible that Feiler's book, *America's Prophet: How the Story of Moses Shaped America*, was a much bigger bestseller than I realised at the time.) But either way, I can't see any actual relevance to the declared aims of the article.

So let's call that the '**Shout about something completely irrelevant'** **argument.**

Feiler's third argument is that there aren't many atheists in America. In the survey, 6 per cent of participants said they didn't believe in God, 1 per cent said they didn't know, 69 per cent were absolutely certain that God exists, and 17 per cent were fairly certain [see box below]. In addition, Feiler argues that, contrary to stereotype, those 69 per cent of believers aren't particularly dogmatic since only a third of them say the Bible should be taken literally.

Incidentally, most atheists I know would regard the idea that 23 per cent of the sample believed in the literal truth of the Bible as proof that Christians in America are pretty dogmatic indeed, but let's put that to one side.

Anyhow, the point of this strand of Feiler's argument is summed up in the headline: 'Believers still dominate in America; atheists are still rare'. So let's call that the '**We win anyway, so there!**' **argument.**

Feiler's fourth and final argument is pretty convoluted. It starts with a slight red herring, pointing out that some articles written about the survey had implied that the survey suggested that Americans were ignorant about other religions, citing the fact that on questions like the religious make up of Indonesia, the identity of Shiva, and describing nirvana, only 25–40 per cent had given correct answers.

To rebut this, Feiler points out that when it came to other answers, two-thirds could identify the predominant religion of India, 70 per cent the predominant religion of Pakistan, and 82 per cent could identify the religion Mother Teresa practised. And to conclude this list, 'amazingly' (in the author's bombastic words) more people knew that Ramadan is an Islamic holy month than could name the author of *Moby Dick*.

I'm going to digress for a moment here. Consider the rhythm of the list above. We start with India at 66 per cent, then go up to 70 per cent and 82 per cent before the pay-off of the comparison of Ramadan and Herman Melville. Humans are good at recognising and projecting patterns, and this rising series naturally leads the mind to expect it to continue. So given that 'amazingly', our intuitive understanding is likely to be that over 82 per cent of participants correctly identified Ramadan.

In fact, when we check the survey figures online we find 52 per cent correct answers for the former question and 42 per cent for the *Moby Dick* question. A cynic might conclude that the *Moby Dick* question was intentionally chosen as one that more people got wrong than right, and that the actual figures aren't included in the article as they would be somewhat bathetic.

Anyhow, Feiler's conclusion is that Americans do indeed know a fair amount about other religions and that, given America's involvement (at the time) in two wars in Muslim countries, this might be a key national security advantage in years to come.

I don't think I can sum that convoluted argument up in a single line: I guess it's a combination of the **smokescreen** (by which I mean any attempt to throw up a confusing array of irrelevant thoughts that obscure the truth) and the **'Raise the flag and play the national anthem' argument**.

Of course, none of this suggests that Feiler is lying. It is almost certainly an example of motivated thinking (see p. 64) in which strong beliefs that are already held get in the way of a calmer, more rational approach to the data. And of course, in the fourth argument above, the appeal to patriotism is likely to invoke a similar state of mind in the susceptible reader: if it is presented as being patriotic to reject the evidence that atheists know more about religion than believers, then many who see themselves as patriotic will be won over.

The irony is that Feiler needn't have gone to such lengths; his third argument above contained the only clue he needed to see what was really wrong with the headlines: the relative scarcity of atheists in the survey. Charles Seife analysed the problem in far more elegant terms, pointing out that firstly, since only 212 out of the 3,412 participants labelled themselves as atheist, the sample for atheists was very small and thus had a large margin of error (see p. 12). Secondly, Pew excluded from the summary those who said they believed in 'nothing in particular', who had also said they didn't believe in God and thus should have been included in the 'atheists' category in spite of not self-identifying that way. Adding these people

(2 per cent of the total sample) to the self-identified atheists (6 per cent of the total sample) went a long way to levelling up the figures as their perform-ance in the test proved to have been below average. Thirdly, in addition, education and income are likely to co-vary with general knowledge, and it is at least likely that atheism will also co-vary with education.

Weighting for those factors would have made the conclusion of the survey on this subject far more accurate but less interesting. 'Atheists And Believers Know Roughly The Same Amount About Religion, But We Didn't Really Ask Enough Of Them To Be Sure' doesn't really make for a good headline, does it?

Bullet Point: Check the actual figures and delve into the factors that haven't been reported on in the headline summary.

MARGIN OF ERROR

Any time you are considering the results of a survey or opinion poll, it's worth taking a moment to consider what is going on when it comes to the margin of error and confidence levels.

Consider a sample from a population that has normal distribution (in other words, the graph of results follows the typical bell curve in which most of the results will be a band in the middle, and in each direction the outliers will gradually decrease towards zero). The first thing to bear in mind is that the bigger the sample we are working from, the more concentrated the results will be around the actual mean of the popula-tion, which gives us a decreasing margin of error.

One rule of thumb for understanding this relies on the concept of **standard deviation**. This measures how widely dispersed the individual elements of a sample are, on average.

To calculate it, we first find the mean of all the results in our sample. Then we calculate the difference of each individual piece of data from the mean. We square these (meaning we have a positive value for every piece of data whether it is above or below the mean) then find the average of the differences, which is the variance. Then we find the square root of the variance, which is the standard deviation.

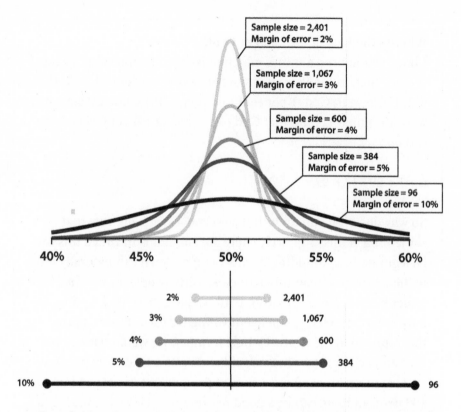

If we have a large sample with normal distribution we can use the 68/95/99.7 rule. This tells us that 68 per cent of the set will fall within one standard deviation of the mean, 95 per cent of the responses will fall within two standard deviations, and 99.7 per cent will fall within three standard deviations.

Now, if a pollster or reporter tells us that for any given survey, the confidence level is 95 per cent and the margin of error is 3 per cent, what this actually means is that we can expect the actual result to be within 3 per cent of the prediction 95 per cent of the time. The actual calculation of margin of error is a bit technical, as it involves calculating critical value (using either a 't-value' or, more often, a 'z-value') and calculating this using standard deviation or the closely related standard error (depending on whether or not we know the standard deviation for the entire population). Forgive me if I don't go into every detail of this: there are some excellent videos explaining the detail online if you want to learn the skills involved.

Where's the Rest of the 100 Per Cent?

I hope you noticed a minor, niggling issue in the previous section. We saw that, in the Pew survey, 6 per cent of participants said they didn't believe in God, 1 per cent said they didn't know, 69 per cent were absolutely certain that God exists and 17 per cent were fairly certain.

$$6 + 1 + 69 + 17 = 92$$

So why doesn't it add up to 100 per cent? Does it indicate that something is missing? Well, there are situations in which this can be legitimate, as rounding numbers up or down can naturally lead to this outcome. Imagine a survey of thirty people on whether they prefer peas, beans or broccoli. If 10 people choose each option, we have a third or 33.33 . . . per cent voting for each. If we reported this rounded to the nearest whole number, we would have 33 per cent for each option, which would add up to a total of 99 per cent.

This means that where a sum of percentages doesn't add up to 100 per cent there can be a good reason for it. However, there is a limit to how much error can be introduced this way, and we can create a mathematical formula to define that limit.

Let's call n the number of rounded terms that are being added up to the total (which we know should be 100 per cent for the unrounded figures). What is the most each can contribute to the rounding? Say we have a number like 25.4 or 37.8. We can call the 0.4 and the 0.8 the incomplete units within the total.

The convention is that when it comes to the incomplete unit in the total we round anything below 0.5 down to 0 and anything equal to or greater than 0.5 up to 1. This is only a convention, so it's best not to assume that for some quirky reason the particular company isn't rounding down from 0.5 to 0. But the one thing we can be sure of is that the maximum error introduced by a single rounded term to the overall total is 0.5.

So the maximum error that rounding can introduce to a sum of n rounded terms is 0.5n. It will usually be significantly less, of course, as there is likely to be rounding up as well as rounding down, so part of the error might cancel itself out. But in the most extreme case every rounded term would have an incomplete unit of exactly 0.5 so 0.5n would be the most extreme error that could be introduced.

Going back to the Pew survey, we had four rounded terms, which could introduce a maximum error to the total of 2 per cent. But the actual error is 8 per cent. So where did that 8 per cent go?

This is where you have to dig into the original data to find the truth.

Luckily the full report is still available. There we find that in addition to the categories quoted in the article, there are three further categories: as well as the category of people who were fairly sure God existed, there were 4 per cent who were 'not too certain', and 1 per cent who were 'not at all certain'. And, as we saw in the previous section, there was also 2 per cent of the sample who had said they didn't believe in God but refused to answer what they believed or otherwise failed to declare themselves an atheist.

So we have an extra $1 + 2 + 4 = 7$ per cent. This brings the total of seven terms up to 99 per cent. Now we have a 1 per cent disparity from 100 per cent, but 7 terms gives us a potential error introduced by the sampling process of $0.5 \times 7 = 3.5$ per cent.

The formula above isn't infallible at indicating when something has been left out, as a small term could be excluded without falling foul of the formula. However, it is a good rule of thumb to use it to raise a red flag where one or more of the categories has been omitted, because a total that is wrong by more than the maximum amount indicated by the formula will always indicate that there is some problem, even if it is just a typo on the behalf of the reporter.

The key thing is that there is an inherent danger involved in the stated confidence levels and margin of error. These will be at the maximum and minimum respectively for a binary question in which the result is close to 50/50 (for instance, if you asked whether someone's favourite colour was blue or green and the results were fairly even).

One problem comes when you have something like a poll that predicts the result of an election with parties of varying levels of popularity. The larger the party's probable share of the vote, the higher the actual confidence level will be, whereas the smaller parties will be based on a smaller sample, so in spite of the fact that they are part of a large poll, we need to apply more scepticism to the results for them (just as Pew should have been more sceptical of the results for the atheists that made up 6 per cent of their total sample in the example above).

At the same time, the actual margin of error for that smaller party might be a smaller numerical value. If this sounds counter-intuitive, consider the fact that a party projected to get 10 per cent of the poll who gets 11 per cent in the actual election will have exceeded the projection by 10 per cent, while a larger party projected to get 50 per cent who actually gets 52 per cent will only have exceeded the prediction by 4 per cent even though the prediction was out by a larger number of votes than it was for the smaller party.

The essential moral here is that, up to a point, the stated margin of error and confidence level is worth paying attention to; but it is only one broad measure that is being applied to the headline findings of the survey and the devil is in the detail.

> **Bullet Point: Do pay attention to confidence levels and margins of error, but bear in mind they come with their own health warnings and can be misleading in their own ways.**

CHERRY-PICKING FOR BEGINNERS

If you want to make a case for a particular proposition and find that the evidence isn't on your side, one way of making your argument is to **cherry-pick** the evidence. Of course, there is a legitimate argument for

cherry-picking. A lawyer making a case for the defence or the prosecu-
tion will inevitably choose those facts that best suit their case, while
ignoring those that are less helpful (while the opposing lawyer will probably be making the opposing choices).

One problem with the modern political system, which often pits
parties or personalities against each other, is how often politicians end
up acting in a similarly adversarial way. Rather than starting from the
facts and proceeding to conclusions, they are increasingly likely to start
with the conclusion and then look for the evidence that suits that case.

Cherry-picking is the inevitable result.

Here are a couple of examples from either side of the US political
divide. In 2019 the White House website made the claim that under
President Trump, 'Economic Growth Has Reached 3 Percent for the First
Time in More than a Decade'.

Now, there was an immediate problem with this claim. The usual way
that economic growth is measured is using the increase in GDP over a
twelve-month period. But growth in the twelve months ending in the second
quarter of 2015 had been 3.4 per cent. For the year ending in the first quarter
of that year growth was 3.8 per cent. And there were two other occasions
while Obama was in office on which growth over a twelve-month period had
been 3 per cent or higher.

So, puzzled economists and journalists had to go back to study the
small print of the White House website. There they discovered that
the White House economists had relied on a very specific restriction:
the claim was merely that this was the first time for a decade that GDP
as measured in the fourth quarter of the year was 3 per cent or more
than a year earlier.

By Trump's standards this is a relatively small bit of mangling the
truth, but it is revealing, because it shows how the spokespeople who are
working on his behalf have gone out to find a specific statistic that suits
a narrative (Trump had been predicting that GDP growth would be '4 per
cent to 6 per cent' under his presidency since he started campaigning)
rather than merely giving the facts.

Another example comes from the 2018 Senate elections. Adverts
made on behalf of Bill Nelson, the Democratic challenger, claimed that

his opponent Bill Scott had taken over a billion dollars out of education in Florida during his time in office. This came with a PolitiFact stamp of approval, which matters as PolitiFact is an organisation that aims to provide some objectivity in the age of spin, and their approval might be taken to mean something.

However, there is a problem here. The PolitiFact stamp of approval was for a claim that Scott had taken $1 billion out of the US education budget in his first year in office, 2011. Charlie Crist, the Democratic challenger in the subsequent gubernatorial election, had made that claim, which was essentially true. But Scott's record over subsequent years had been to restore that funding to the point where both overall and per-pupil spending were higher at the end of his tenure than they had been at the start. So when PolitiFact were asked to recheck the claim in 2018 they were obliged to give it a rating of 'mostly false'.

Obviously the maths of these examples is pretty basic, but the more complicated the maths and the statistics become, the greater the chance that ordinary people can be hoodwinked by grand-sounding claims that might essentially be true, but have been carefully cherry-picked to make a case, whether it is deployed in the argument over climate change and what to do about it or in any other political sphere.

> **Bullet Point: If a politician uses a statistic, ask why they chose that particular statistic and investigate the ones they chose not to use.**

When Numbers Drive Actions

Political promises are often couched in terms of numbers. 'We will get immigration down to the tens of thousands per year'; 'We will get unemployment down'; 'We will drive economic prosperity and growth'; 'We will get the crime figures down'. And the role of government has increasingly focused on the setting of targets that are intended to make those promises come true. I want to briefly focus on a few of the ways in which this tendency can backfire in real life.

Debt or Deficit

Sometimes it is hard to tell whether the politicians are numerically challenged or whether they are assuming that the public is and are relying on that to deliberately distort the debate. For instance, consider the many ways in which 'debt' and 'deficit' get confused in the public discourse.

In the period after the global financial crisis, various governments around the world promised to use 'austerity' (in other words cuts to government expenditure) to address the problem of increased national debt (which was caused by the enforced bailout of the banking sector).

In the UK, there were numerous examples of politicians mixing up debt (which means the total amount owed) and deficit (which means the difference between what the government is spending and what it is receiving). At one point the Centre for Policy Research commissioned a report that showed that only 10 per cent of the British public knew that the government's austerity package was likely to increase national debt by £600 billion over the course of the parliament, whereas 47 per cent believed the precise opposite. The report's author, Ryan Bourne, pointed out the government's true aim was much less ambitious, as they were promising that the 'ratio of debt to GDP' would be falling by the end of the parliament (less ambitious, but even more mathematically confusing). He said: 'The Government has often said that it wants to pay the nation's credit card off. But if it wants to use that analogy correctly, it should be saying that it wants to reduce the amount that is added to the credit card debt each year.'

There is another debate to be had about this, which is whether it was ever that crucial to reduce the national debt so rapidly, and whether the government's focus on austerity allowed the bankers to escape the blame for their central role in the crisis. I will return to this in the finance chapter, but for now, the main point is that when politicians talk about debt and deficit, you need to listen very carefully to what is actually being said.

Firstly, government policy can itself be distorted by the aim of having the maximum possible impact on the statistic that is being measured. To explain this, here's a brief summary of some government policies in the UK regarding the unemployment figures, which have long been a source of controversy, especially since they peaked in the early 1980s after a long rise at over 3 million. I'll explain this in broad terms, as I know there has been a similar pattern of government policy in many other countries (although the exact details vary).

In the early 1970s, invalidity benefit was introduced for the first time. This meant that rather than simply being on the same unemployment benefit as fit people without a job, those who had a long-term injury or illness were treated differently (and, in some measures, were excluded from the unemployment figures).

In the 1980s, a 'National Enterprise Scheme' was created: this allowed people who were unemployed to register themselves as 'self-employed' in order to set up and develop a business. The main attraction to claimants was that for that period of time there was no need to report to the unemployment office to 'sign on' and they received the same amount of money although there was little attempt to monitor whether or not they really were developing a business. Other similar schemes followed over the years. The outcome of this was that significant numbers of people were temporarily taken out of the 'unemployed' figures, since they were technically in employment.

In the 1990s, invalidity benefit became incapacity benefit. While the unemployment figures excluded those on incapacity benefit, the number of people on it had steadily grown and now became a political controversy in its own right. The result was a series of attempts to 'crack down' on claimants by making them attend 'fit for work' interviews and prove that they were making efforts to get work.

By 2010, the employment landscape had changed significantly: zero hours contracts were steadily becoming a larger and larger part of the economy. At this point, there was an increase in pressure from politicians for both the unemployed and incapacitated to be encouraged 'back to work'. Once again, self-employment was used as a smokescreen. For instance, someone who had been 'encouraged' to take up a zero hours contract (often through the use of sanctions and suspensions of their benefits) would be counted as 'in employment' whether or not they were actually paid in any given week. Oh, and if they declined to take up that zero hours contract for any reason, they could also be excluded from the unemployment figures by the simple method of suspending their benefits.

It's notable that by this stage of the process, government pronouncements have shifted from promises to 'get unemployment down' to boasts that there are 'more people in employment than ever before'. It may not be a coincidence that at the same time there is a record number of people who are in employment but who are forced to rely on food banks (where free supplies are voluntarily contributed by charities to help people in poverty).

The overview is that many of these developments, some of which had deleterious effects on the individual citizens involved, were strongly influenced by the desire to affect the unemployment or employment figures without actually doing the things that would have had the most impact on real communities: creating stable, well-paid jobs, or encouraging employers to do so. And when so much of government policy is focused on the goal of massaging the statistics by changing the way things are measured, you can be sure that the real issues are not being addressed, whether the issue is climate change, employment, crime, education, health or whatever.

Bullet Point: Why would a government attempt to change the real world, when it is so much easier to change the way they measure it?

US Unemployment Figures

I've focused above on the unemployment statistics in the UK. However, it's worth reiterating that many countries have experienced similar issues. I'm not a huge fan of the Shadowstats website on which the economist known as John Williams publishes his ongoing criticisms of US government statistics. However, let's acknowledge his critique of the Bureau of Labor Statistics (BLS) unemployment numbers as a comparison point.

Williams points to the fact that under President Lyndon B. Johnson, the U-3 unemployment rate series was created. This was a stat that didn't include people who stopped looking for work over a year ago or part-time workers who were looking for full-time employment. The previously used U-6 rate continued to be published, but media and BLS sources tend to focus on the U-3 series. Williams regularly calculates and publishes the U-6 rate as it would have been calculated before method-ology changes in December 1993. His argument is that the BLS is deliberately understating the true unemployment rate.

I'll comment more on this when we discuss inflation measurement in Chapter 7, 'What's Wrong with Economics'; for now, it's worth noting that each country tends to have its own specific controversies over meas-urements of inflation, unemployment and economic growth and these will often be seen by some as cherry-picking if not outright distortion.

Bullet Point: Find out how your government calculates the figures and what difference alternative approaches would make.

What Would Hitler Do?
I'll try not to mention the Nazis too often in this book, but it's interesting to note that Hitler made a specific promise that helped bring him to power: against the backdrop of mass unemployment in the Depression of the early 1930s, he promised to return Germany to full employment. By 1939 he was able to claim that he had fulfilled this pledge. How? Well, firstly, his government intro-duced the National Labour Service: all young men had to serve six months in it, following which they were conscripted into military service. The 1.4 million men who were in the army by 1939 weren't counted in the unemployment figures. Neither were women, who were in many cases encouraged or obliged to give up their jobs in favour of men. And nor were the many Jews who had lost either their businesses or their jobs (which had been given to other people). So the claim to have abolished unemployment rested on a very peculiar and unpleasant way of measuring the situation.

FIXING THE OUTCOMES

Another problem with target-driven public policy comes when the people who are charged with carrying out that policy have the means to influence what is included in the statistics.

For instance, in recent years police officers in London's Met force were more likely to be promoted if they were meeting targets. This obviously incentivised them to try to meet those targets. But how can you do that when the targets are based on arrest rates and crime rates? Surely those are down to how many criminals there are and how many crimes they are committing?

In 2013, a British parliamentary investigation found evidence to the opposite. And bear in mind that this is just one example of the sorts of things that happen in any organisation where the participants become too focused on the numbers rather than on the real-world application of those numbers.

Firstly, rape and sexual offences were being routinely under-reported, as those who reported such offences were being discouraged from taking it further. (These are crimes with a notoriously low rate of conviction, which looks bad in the statistics). In many cases 'no crime' was recorded, and there was a tendency towards victim blaming, in which mental health or other 'undisclosed issues' were being put forward as reasons for the crime not being recorded.

Thefts weren't being recorded consistently for the same reason: that they added to the numbers of crimes reported, but had a low rate of conviction or detection. One way that burglary rates were being suppressed was by categorising them in lower brackets, such as criminal damage or 'attempted burglary': this was leading to as many as 150 burglaries a week not being properly recorded. There was also a tendency for multiple burglaries in an area to be written down as a single crime.

At the time, the Mayor's Office for Policing and Crime had set a target of a 20 per cent crime reduction. As the former Detective Chief Superintendent Peter Barron put it: 'When targets are set by offices such as the Mayor's Office for Policing and Crime, what they think they are asking for are 20 per cent fewer victims. That translates into "record 20 per cent fewer crimes" as far as . . . senior officers are concerned.'

Apologies to the Met for continuing to dwell on this one example, but one of the joys of the report is that the police involved had even developed their own jargon for ways of under-reporting crime: 'cuffing', 'skewing' and 'nodding'. Cuffing referred to officers reducing crime by refusing to believe complainants, by recording a theft as 'lost property', for instance, or by recording multiple break-ins as just one crime. Skewing is the practice of putting extra resources into particular areas that are being measured (whether that was the highest crime area or not). And nodding is where there is a tacit agreement between the police and an offender whereby they are not charged for a more serious offence if they allow the police to get an easy conviction for a lesser offence.

By comparison, the old practice of stitching someone up by faking evidence or confessions seems quite quaint.

Anyhow, the main point of all this is that if you ask someone to meet a particular measure as a way of trying to get them to perform their job, then they may well find ways to meet that target that don't match up to the intentions of the target setter.

Another example comes from the British National Health Service. In the early years of the twenty-first century, waits of twelve hours or more were not uncommon in Accident and Emergency (A&E) departments. The government's solution was to set a target: the maximum waiting time at an A&E department should be four hours. In some cases (though not all) extra resources were provided (which, of course, points to the underlying problem). But in many areas this didn't happen.

In spite of this, the problem did appear to improve considerably. But there were a variety of workarounds that were reported as having been used by hospital managers to game the system.

In some cases new wards were set up adjacent to A&E departments and given titles such as 'medical assessment wards' or 'clinical decision wards'. Once the patient was transferred into these, the clock was stopped on the measurement of their time in A&E, even if they spent many hours in what was essentially a glorified waiting area.

In other cases hospital managers refused to admit more patients than they could manage, wherever this was clinically possible. This led in some instances to ambulance crews being asked to wait outside the hospital with patients who had serious but non-life-threatening conditions; and again this could be for hours. The logic was that the patient had only been admitted to A&E once the ambulance crew handed them over, so the slower that was, the less time they were recorded as having waited for. The unfortunate side effect of this was that a significant number of ambulances were being kept off the streets and thus waiting times for ambulances started to increase in many areas.

Just as with the police example above, the managers were only being logical in their behaviour. They had been incentivised to meet a given target, and they were ingenious and resourceful in the ways they tried to achieve that. But the outcome for patients and victims of crimes was often either neutral or negative.

But as those managers might say, the numbers don't lie. Do they?

Bullet Point: If you set someone a target, they might meet it in ways you hadn't predicted.

Measure for Measure

The practice of **scientific management** became widespread from the early part of the twentieth century (when it was propagated by Frederick Winslow Taylor in particular), and ubiquitous in the second half of that century. The hugely influential post-war management guru Peter Drucker once said: 'What gets measured gets improved.' And it can be argued that this has become one of the core beliefs of politicians in recent decades. The belief is that rather than making large-scale changes to the economy and public services, target culture (and 'nudge' strategies) can be used to induce the desired improvements in the economy and society as a whole. And the same is true of many companies, who see measurement as the linchpin of effectiveness management. The idea is that you can lie, dissemble and fool your employer into thinking you're doing a great job, but the numbers can't. So next time you find yourself discussing your key performance indicators (KPIs) in that regular progress meeting, you may want to mutter a curse in Peter Drucker's direction.

TARGETS IN EDUCATION

In America, both the No Child Left Behind programme (introduced in 2002) and the Every Student Succeeds Act that replaced it in 2015 are examples of how target culture operates in education. There are similar systems in many education systems around the world.

The fundamental feature these systems all share is that they rely on the use of targets, which in turn are largely based on standardised testing in core subjects: literacy and numeracy, in particular. In the case of the American programmes, funding may be partly dependent on the outcomes of these tests, and bonuses, school rankings and ratings for individual teachers will also depend on them.

So how does this affect the teachers? Firstly, bear in mind that a teacher can't choose their students. They may have the brightest, best behaved class in the school one year and a much more challenging one the next. They may have more students with special needs or fewer. They may have some exceptionally gifted students in a class. One can imagine that every single class presents different challenges and requires great flexibility and imagination in approach.

The problem is that standardised testing puts the teacher in something of a straitjacket. They have to put as much energy as possible into ensuring their students are trained to pass the tests (which in turn often results in stress for both teachers and pupils). And the feat that will gain most kudos will generally be to get as many pupils as possible up to a particular benchmark.

If the handful of most challenged pupils are well taught by the teacher, and improve considerably but still don't achieve a pass, then those efforts will not be counted. Similarly, where an exceptional pupil is already quite capable of passing the tests, the teacher will not be given credit for challenging them to greater heights. They would have passed anyway. So the teacher is effectively forced to concentrate as much energy as possible on the middle of the bell curve: those middling students whose ability to pass the tests is in doubt. Getting the mediocre students up to a pass grade and making sure that the students who are just about capable of passing don't slip backwards and fail is the most effective way of maximising the test results.

Of course, there is a degree to which this has been true for as long as exam results have been used to measure the final outcome of the educational process. But the tests in the American programmes (and equivalents such as the SATS tests in the UK) are applied from early years and then repeated throughout schooling. So the entire education process becomes more homogenised and focused on the core subjects. Headmasters might get respect and nice mentions in school inspectors' reports for encouraging other subjects – such as art, sports, or fringe subjects – but this will not show up in the all important tests, so there is less incentive to provide a more complex, rounded education. And there is less incentive to focus on the individuality of each pupil: they are instead treated as cogs in a machine.

And what about those inspirational teachers we see in books and films like Mr Chips or Miss Jean Brodie, or the Robin Williams character in the movie *Dead Poets Society*, who famously urged his pupils to *carpe diem* and make their lives extraordinary? Well, it is notable that retention rates in the teaching profession have rarely been lower. The teachers who dreamed of treating each of their charges as unique individuals and encouraging their development spend their lives filling in progress reports and attending constant performance reviews. Of course, any teacher will recognise that those initial dreams are often ground down by the difficulty of real life in the classroom. But the more the most capable teachers come to recognise that they, like their students, have become numbers on a spreadsheet, the less likely they are to be recognised as good teachers or to stay in the profession at all.

Bullet Point: When it comes to education, numbers are important, but actual students and teachers are more important.

How to Improve on Perfection

One of the problems that dogs testing schemes in education is the dilemma over whether the rewards (whether it be in funds, rankings or recognition) should go to schools that deliver 'improvement' or 'excellence'. The former can be achieved by turning a terrible school into a

Rewards and Effort

There have been numerous psychological studies that show that monetary rewards in particular can actually disincentivise people. In one study on primary school children, infants were given drawing equipment and promised a shiny certificate of achievement if they did well. Another group was given the same appealing equipment but no promise of reward. Both groups worked hard to produce nice drawings. The experiment was repeated two weeks later, with neither group being promised a reward. The first group made less effort now the reward had been withdrawn, while the latter group continued to make a strong effort.

In another study, a large group of subjects were asked to donate blood, with one group being offered $10 while the other group weren't: there were more takers among the unrewarded group. Similarly, a group of lawyers were asked to provide legal help to a needy case for a reduced fee while another group were asked to do so for free. Again, the latter group were more willing to help. In both cases, the lack of reward meant that their attention was focused on different motivational factors, such as their altruism or compassion, while the use of monetary reward made that group focus on whether or not the reward was sufficient.

In general, the studies show that material rewards reduce the attraction of activities that are pleasurable, and that any incentive they provide for doing less pleasurable activities halts as soon as the reward is withdrawn. It can be argued that this casts doubt on the entire psychological effect of measurement, targets and rewards in the workplace.

But that is unlikely to make any difference to the pervasive target culture of our times.

mediocre one. But if an excellent school slips for a few years into merely being very good, is it fair for them to be penalised? And while we can recognise that a teacher who delivers improvement in each individual pupil in their care is doing an excellent job, it is rare for the benchmarks to be finely calibrated enough to measure this; instead, they may well be castigated for having a slightly more difficult class to teach than in the previous year.

This brings to mind an anecdote from my wife's time at primary school, aged about eight. In the first maths lesson of the year she was given a test: having always found maths easy, she got every question right, and was pleased to get a perfect mark. The following week, she again got 100 per cent in the test. As the teacher read out the marks, they awarded 'merits' to every pupil who had beaten their previous score. And this process continued throughout the school year. The students whose marks were erratic were rewarded every time their marks bounced up. As my wife was far too conscientious or proud to deliberately fail any of the tests, she continued to get top marks.

And as a result, to her ongoing chagrin, no 'merits' were forthcoming.

Bullet Point: Targeting improvement in performance rather than performance can lead to counter-intuitive outcomes.

TURNING HUMANS INTO MACHINES

The problems associated with targets and key performance indictors (KPIs) aren't only restricted to situations in which machines are actually carrying out the algorithms that process the information. A large part of the administrative effort of any organisation that succumbs to target culture is about monitoring performance and 'compliance' (such a sinister word).

As this is built into the way the organisation operates, it increasingly requires ordinary people doing their jobs to behave as though they are machines carrying out algorithms. The excellent book *Tickbox* by David Boyle focuses on tickbox or checkbox culture, and the way it obliges workers to behave. One of the most depressing examples comes from the way that the UK government dealt with the problems in the immigration service that have come to be collectively known as the Windrush scandal. This involved people (especially those of Caribbean origin from

Reduced to a Five-Point Scale

The five-point consumer satisfaction survey is one particular way of reducing interactions with customers (or citizens, in the case of government departments) to an easy set of information that can be input to a programme. In general you will be asked a series of questions, designed by the company or organisation, for each of which you tick one of five boxes ranging from Very Unsatisfied to Very Satisfied. There is, of course, some leeway for the organisation to choose the questions: the more the question can steer you to focus on the customer operative you talked to (who may have been quite helpful in communicating the organisation's unhelpful rules and is at least a real person you just interacted with) rather than on the organisation's behaviour as a whole, the better the results will be. You do often get the opportunity to add 'Any Additional Comments' in a box at the end, so you aren't entirely being reduced to numbers, are you? Well, I know from personal conversations that those additional comments are not always logged in the system; they are in some cases only there to allow you to feel you have said what you really want to say, while the only thing that will show up in the company's figures will be the overall satisfaction ratings. This is a pretty reductionist way to treat people, even at face value. It may make you feel slightly more queasy about the process to know that the system was invented by the American sociologist Rensis Likert, and was first used in an extensive way when the American authorities asked Japanese and German civilians after the war how they had felt about being bombed during it.

Unsurprisingly, they weren't very satisfied with the experience, as a whole.

the 'Windrush generation', some of the post-war wave of immigration that had been encouraged by the UK government) who were detained wrongly, denied their correct legal rights, threatened with potential deportation, and even wrongly deported from the UK in eighty-three notable cases. These were people who had arrived in the country before 1973, so had been living and working here for many years, and who were often part of families who were raised entirely in Britain.

When these people first arrived they had the right to live in this country, and didn't need any documentation to prove this. Subsequently, the immigration laws changed, but they were still given no proof of their ongoing right to residency. The problems arose when, for a variety of reasons, their right to be here was eventually challenged (in some cases after they had returned to the Caribbean for a long-awaited family reunion).

Boyle discusses the way that tickbox culture led Home Office employees to be incentivised to deal with the easy applications when it came to immigration and asylum applications. Where there were discrepancies, or complications, the cases tended to go to the bottom of the pile as departments attempted to meet their targets. The result was that many of those caught up in the anomaly were left in a horrible no man's land for years: some of them died while still waiting for their right to reside in the country where they had lived and worked for most of their lives to be confirmed.

This is only one example, but it is a potent warning of what can happen if humans are asked to meet targets and to carry out limited form-filling and box-ticking rather than treating each case that comes before them as an individual case of a person in need of help or service.

Bullet Point: If organisations operate like machines, and employees are mere cogs in that machine, then the machine may output some pretty intolerable results.

Don't Mention the War

OK, I mentioned the Nazis once, but I think I got away with it all right . . . But now I am going to risk succumbing to Godwin's law once again. (Technically, Godwin's law applies to internet forums and discussions, and states that, as a political debate progresses over time, the probability of someone invoking a comparison with Hitler or the Nazis approaches 100 per cent.)

Goodhart's Law

It's probably fairly clear what I think about target culture and scientific management by now. It's worth noting that various sociologists, economists and business commentators have given formal statements of why targets tend to go wrong. In the 1970s, Donald T. Campbell formulated **Campbell's law**: 'The more any quantitative social indicator is used for social decision-making, the more subject it will be to corruption pressures and the more apt it will be to distort and corrupt the social processes it is intended to monitor.' In a similar vein, the **Lucas critique** (named after Robert Lucas) states that: 'Given that the structure of an econometric model consists of optimal decision rules of economic agents, and that optimal decision rules vary systematically with changes in the structure of series relevant to the decision maker, it follows that any change in policy will systematically alter the structure of econometric models.' These are pretty fair statements, but quite wordy. The economist Charles Goodhart managed a slightly more succinct version: 'Any observed statistical regularity will tend to collapse once pressure is placed upon it for control purposes.' But finally, the anthropologist Marilyn Strathern really nailed this down in her restatement of what has come to be known as **Goodhart's law**: 'When a measure becomes a target, it ceases to be a good measure.'

A lot of organisations could do with hanging that on a big sign over the entrance.

The Nazi regime in Germany was highly focused on the use of statistics and propaganda. Friedrich Zahn, president of the Bavarian Statistical Office, proudly noted in *Allgemeines Statistisches Archiv* (ASA), the journal of the German Statistical Society, that: 'The government of our Führer and Reichschancellor Adolf Hitler is statistics-friendly.'

To get a feel for Hitler's attitude to propaganda, consider the quote below from *Mein Kampf*. In the book, he spent two chapters specifically discussing propaganda, an obsession that arose out of the feeling that the German government had been ineffectual in its response to British propaganda in the First World War:

> The function of propaganda does not lie in the scientific training of the individual, but in calling the masses' attention to certain facts, processes, necessities, etc., whose significance is thus for the first time placed within their field of vision. The whole art consists in doing this so skilfully that everyone will be convinced that the fact is real, the process necessary, the necessity correct, etc. . . . its effect for the most part must be aimed at the emotions and only to a very limited degree at the so-called intellect.

He continued to insist that:

> . . . all propaganda must be popular and its intellectual level must be adjusted to the most limited intelligence among those it is addressed to . . . to influence a whole people, we must avoid excessive intellectual demands on our public, and too much caution cannot be exerted in this direction . . . all effective propaganda must be limited to a very few points and must harp on these slogans until the last member of the public understands what you want him to understand by your slogan.

When it comes to modern political slogans such as 'Make America Great Again', 'Change We Can Believe In' or 'Get Brexit Done', one can perhaps see a modern reflection of this ruthlessly simplistic approach.

During the Second World War, statistics became a part of the battle. For instance, one German statement from 1940 was called 'Churchill's False Figures': it was a direct response to British reports that their total losses in October that year were 21,867 men, and 1,170 in German POW camps: 'The truth, however, is that at that point German POW camps held 1,550 British officers, and 35,500 NCOs and soldiers, or a total of 37,050 English prisoners. That shows the nature of Churchill's "truths".'

Similarly, reports from the *Reichsministerium für Volksaufklarung und Propaganda* (Pro-Mi) during the Battle of Britain focused on emphasising

German superiority: for instance, on 9 August 1940, Pro-Mi claimed that forty-nine British planes had been destroyed as opposed to ten German ones, whereas the BBC counterclaimed that the true numbers were fifty-three German planes destroyed to just sixteen British. Joseph Goebbels once pointed out with satisfaction that air battles were easy to misreport, as, amid the confusion of the skies, there was much confusion that could be taken advantage of.

Late in her life, Brunhilde Pomsel, Goebbels' secretary, recalled her time at Pro-Mi. Among the tasks allotted to her was the massaging of statistics: reporting fewer soldiers lost and exaggerating the number of rapes committed on German women by the Red Army. (Fear of the latter was fanned to such a degree that many German women close to the front line took their own lives as the Russians approached rather than risk falling victim to the brutality described in the propaganda.) She prosaically described all this as 'just another job' and noted that Goebbels and his wife were always nice to her.

Statistics also played a major role in the Holocaust. Statistical publications with titles such as 'Development of German Population Statistics through Genetic-Biological Stock-Taking' and 'The Position of Statistics in the New Reich' give a dully horrifying glimpse of the mindset of the period. Prof. Dr Johannes Müller wrote in a 1934 edition of ASA: 'Above all, remember that several very important problems are being tackled currently, problems of an ideological nature. One of those problems is race politics, and this problem must be viewed in a statistical light.'

IBM had already subsumed the Tabulating Machine Company, which had developed information-storing technology from the late nineteenth century, using punch cards (which were first used in Jacquard looms a century earlier, to replicate patterns in material). Now the systems developed by the German IBM subsidiary, Dehomag, relied on punch cards to automatically process and store large amounts of data.

Historically, census information has had an unfortunately close relationship with genocide and discrimination (something that has been seen more recently in Rwanda, in particular). Before the Nazi invasion, a 'comprehensive population registration system for administrative and statistical purposes' was in use in the Netherlands. After the invasion, the system was updated to identify Jewish and Roma individuals. This was one of the reasons why the fatality rates among targeted groups was so high in the country: 73 per cent of Dutch Jews died as opposed to the much lower rate of 40 per cent in Belgium.

One accidental outcome of the Nazi regime's meticulous approach to statistics and data was that there was a significant amount of documentation captured after the war that helped (in some cases) to identify the victims. However, even in the documents there was a significant degree of deception: camp records from Auschwitz commonly gave plausible causes of death rather than the truth. For instance, when it came to the two young prisoners Mieczysław Rycaj and Tadeusz Rycyk, who were killed by phenol injection in 1943, the causes of death were recorded as bilateral pneumonia and septic pharyngitis respectively.

In addition, when mass killings were carried out, the dates of death would be spread out in the records to further muddy the water. However, in spite of these kinds of deception, the true victims' names were often recorded, and while many of the documents were destroyed late in the war, the ones that were seized helped convict the guilty men at the Nuremberg trials.

Bullet Point: Statistics aren't evil but they can be put to evil purposes.

MEASURING GENOCIDE

While we are on the subject of murderous regimes, it's interesting to note a more positive use of statistics. Patrick Heuveline, a professor of sociology at UCLA, recently spent six years researching the extent of the death toll in Cambodia (as Kampuchea was renamed after the civil war) under the Khmer Rouge, led by Pol Pot.

Pol Pot's regime was a peculiarly paranoid and savage one. He had a particular hatred for the urban, educated intelligentsia, many of whom were either killed or sent to labour in the fields in terrible conditions. Early on, the Khmer Rouge leadership announced on state radio that in order to build the agrarian socialist utopia they were aiming for, they only

---

Soviet Statistics

Of course, the Nazis weren't the only dictatorial regime that relied heavily on statistics. From the 1920s onwards, the leaders of the Soviet Union were attempting to industrialise the economy of the country at a rapid pace. This was rooted in the 'Five Year Plans', the first of which ran from 1928 to 1932 (finishing ahead of schedule): workers were exhorted to work harder with slogans such as, 'Plan is law, fulfillment is duty, over-fulfillment is honour!' Ambitious targets were set for the numbers of factories built, and for various specific targets like the number of tractors built. The drive to meet these targets was treated as a new 'front' following the wars that had preceded them. Officially the first five-year plan's targets were 93.7 per cent achieved in just over four years (at which point a new plan was put in place), although it has often been noted that the Soviets' own statistics were hugely unreliable as they were used as propaganda. In any case, there were other significant problems: the targets were often met by cutting corners on quality and many of the items produced were in fact never used. And the excessively rapid collectivisation of farms led to severe food shortages: the famine of 1932–3 killed somewhere between 3 and 7 million people and left many more disabled. Once again, this is an (especially tragic) example of how a focus on targets can lead to serious problems when the steps taken to 'meet the target' override all other considerations.

---

needed 1 or 2 million citizens: to the rest of the 10 million citizens they coldly said that 'To keep you is no benefit, to destroy you is no loss.'

On fairly arbitrary grounds, the regime categorised citizens into three groups: those with full rights, those who might qualify for full rights later, and 'depositees' who had no rights. These were mainly the residents of the cities, which became ghost towns after the Khmer Rouge force marched the population into the countryside. Depositees were given a ration of two bowls of rice soup a day, and many died of starvation. Meanwhile, there was widespread torture, many executions, and the horrifying deaths on the 'Killing Fields'.

When a dissident group of communists, assisted by Vietnamese forces, overthrew Pol Pot's regime, they initially estimated that there had been 3.3 million deaths. This was based on a house-to-house survey, but as they included the number of victims found in the mass graves in the countryside, it was almost certainly an overestimate, because of the double counting involved.

There were various attempts to come up with a more accurate estimate of the true death toll. More recently, Heuveline attempted to collate all the available data sources and come up with the most accurate estimate possible.

Previous estimates had allowed for factors that would have affected Cambodia's population had the Pol Pot regime not occurred, such as pre-war census figures, fertility rates and the average age, while also making adjustment for the prevalence of death from other sources. But while those studies had been based on a handful of factors, Heuveline used forty-seven different pieces of data, including permutations of migration and the way in which the adoption of birth control was projected to have affected the birth rate.

In all, Heuveline took into account 10,000 combinations of the variables, and ran them through computer simulations, which revealed that there is a 95 per cent chance that the death toll was between 1.2 and 2.8 million. (The frequency with which the simulations produced a specific death toll increased the probability that the number was correct.) And within this, he found that there was a 70 per cent chance it was between 1.5 and 2.25 million. His core projection was that there had been 2.52 million excess deaths, of which 1.4 million were the result of violence.

Why is this so important?

Heuveline explains it thus: 'If the death toll appears untrustworthy, people are more likely to question the extent of the evil that occurred. The trustworthiness of the scale is important because the scale was so massive.'

We've seen over the last sixty-five years the ways in which Holocaust denial can take hold. So there is an absolute historical imperative that we do our best to keep accurate figures of such atrocities to minimise the chances of people lying about or questioning the numbers in future.

Bullet Point: The numbers don't always lie.

Conclusion: Politicians, Numbers and People

Of course, not all politicians end up as genocidal lunatics. Many genuinely want to do the right thing, and some of them even end up succeeding. But there are huge dangers involved when politicians become too dependent on targets, measures and statistics. Firstly, this may tempt them into targeting the figures rather than reality. Secondly, it will only encourage their natural propensity to lie, to distort the figures to make it look as though they have fulfilled their promises. All of this is a distraction from the job they are supposed to be doing, so no matter how fruitless it may sometimes seem, it is crucial for us to take stock of how accurately they are using the figures, to scrutinise their methods and choices and to hold them to the highest possible standards of honesty in the way they report data.

In the next chapter, we will go back to basics and consider some of the ways that people abuse data and statistics, even when they are not entirely making them up.

In the classic television series *The Prisoner*, the central character finds himself marooned in a remote coastal village, having been abducted. He is given the new name of 'Number Six' and is never referred to by his real name, leading to his famous shout of, 'I am not a number, I am a free man!' Most of us who have suffered at the hands of governmental bureaucracy at some point or another will sympathise.

If a politician starts using numbers and statistics, it is always worth bearing this in mind: they are supposed to be acting on behalf of you and your fellow citizens. Always ask them, 'What does this mean in terms of actual individual people?' Don't accept the numbers as being the whole truth, or even part of the truth unless they can first give a clear answer to that simple question.

Bullet Point: You are not a number. You are a real person.

CHAPTER 2

Lying for Beginners

....

Let's take a look at some of the basics of how to use statistics misleadingly. Some of these may be a bit obvious, but it's worth reviewing what can go wrong when people use numbers to make a case.

BLUR THE BOUNDARIES BETWEEN CORRELATION AND CAUSATION

If two events (let's call them A and B) are correlated, it means that they move in the same direction over time. So if A increases, so does B and vice versa. There are four possible explanations.

1. A causes B
2. B causes A
3. A third factor, C, causes both A and B
4. There is no causal connection between A and B whatsoever

Actually, this isn't an exhaustive list as there are also more complex possibilities, such as 'A causes C, which causes B', which is indirect causation rather than direct causation, but for our purposes the simplest four categories are sufficient.

The fourth explanation might be the case if, for instance, the time period in question is limited and only covers a period in which A and B were correlated by pure coincidence, so the first lesson here is that we can't even start to infer causation if we are looking at too small a time period.

The third explanation is also always worth watching out for; for instance, consider the claim that baldness causes cardiovascular disease. The two phenomena may well be correlated, but it is far more likely that they are both caused by a third party: old age, which increases the chances of both significantly.

In general, when we find a case where A and B are correlated, we have a tendency to assume that one of the first two is the explanation. Tyler Vigen's excellent book and website *Spurious Correlations* documents an amusing variety of cases in which two phenomena that obviously aren't related are nonetheless correlated. For instance, over the period 1999–2008 the age of Miss America correlated pretty closely with murders by steam, vapours or hot objects. Over a similar period, US crude oil imports from Norway correlated with the number of drivers killed by a train.

In these cases we find it easy to see that the correlations are coincidental, because we can't imagine a connection. But the human brain is naturally programmed to look for patterns, so when there are two phenomena for which we can imagine a causal link, then we are far too prone to believing that there really is one.

How can liars exploit this tendency? Well, think about the famous 1946 Camel advertisement, which claimed that: 'More doctors smoke Camels than any other cigarette.' At a time when people were starting to realise that smoking might possibly not be very good for you, this exploited people's natural belief that doctors know what is healthy, so they would only smoke the cigarette that was the healthiest one. The advert thus implied a causal link between 'knowing what is healthy' and 'smoking Camels'.

Now, firstly, there is absolutely no reason to believe that doctors' knowledge of health leads them to make healthy choices themselves. It is possibly true that taking drugs such as amphetamines is more common among doctors than the general population, but if it is true it would more likely be because they work such long, stressful hours than because they truly believe amphetamines are health-giving.

As it happens, the deception was twofold in this case. The research on which the claim was based had been carried out by the tobacco company R. J. Reynolds' ad agency, William Esty & Co. Their methodology wasn't terribly sound: the staff carried out interviews with doctors either at medical conferences or at their surgeries, having started the interview by giving them a present of a carton of Camels.

However, even if the claim had been based on solid research rather than such a biased methodology, there would have been no reason to believe that the correlation genuinely implied causation.

Bullet Point: If someone presents you with a correlation, look for confounding factors and don't jump to assuming causation.

PRETEND THAT CAUSATION IS A PERMANENT STATE OF AFFAIRS

Imagine you are working at a gadget company that manufactures food mixers. One year the sales director suggests manufacturing upmarket coffee makers. The launch goes fantastically well, and the company's turnover increases by 100 per cent in the first year and another 100 per cent in the next year. As a result, the sales director makes a presentation at the company meeting. He quite reasonably points out that increasing coffee maker production has led to an increase in profits and shows you a graph that projects growth in turnover of 100 per cent for the next two years.

I assume you would be sceptical, and quite rightly so. The change in focus may well have caused a spurt of growth in turnover, but there will come a point where the market becomes more difficult as you saturate your initial target market and other manufacturers react with competitive products. As an example, think of Nintendo's situation in 1990: at that stage the talisman of Super Mario had helped them to reach a dominant position, with over 90 per cent of the video games market in the USA. But just a few years later, Sega had leapfrogged them into the number-one position following the launch of Sonic the Hedgehog and an expanded range of games. A run of success at a company can't be guaranteed to continue, especially if it fosters a sense of complacency.

A similar phenomenon can be observed in the natural environment. Increased growth of berries in a forest might lead to a growth in the insect population, which might in turn cause a growth in the population of the bird species that feed on them. However, as the population grows, it faces different challenges: for instance, if a disease arrives, the population may be more vulnerable to an outbreak, or a predator may arrive that feeds on that particular bird species, drawn to the forest by their prevalence. The indirect causal link between the berry crop and the bird population was real, but it can't be relied on to last indefinitely.

Finally, consider the relationship between income and happiness: as income rises for a relatively poor person, it tends to have a major impact on their way of life and general happiness. But there comes a point where they have enough income to be comfortable, and thereafter any further increase in income will have **diminishing returns.** Once again, a genuine case of correlation and causation only holds within a certain range, and beyond that it breaks down.

Bullet Point: Correlation may indicate causation, but you shouldn't count on the link being a permanent one.

Don't Define What Average You Are Using

If you have a data set, there are at least three ways of calculating the average value. You probably learned this at school, but I'll remind you of the basics.

To calculate the mean, you add up the values and divide by the number of instances. To calculate the median, you put the values in order from smallest to largest, and find the one in the middle. To find the mode, you find the value that occurs most often.

So, consider this set of values for the income of fifteen people living in a particularly diverse street.

£1,000,000
£150,000
£50,000
£50,000
£40,000
£40,000
£40,000
£20,000
£20,000

£19,000
£18,000
£17,000
£16,000
£15,000
£5,000

The sum is £1,500,000. So the mean income on this street is £100,000. The median is £20,000 and the mode is £40,000. This gives a potential liar a massive advantage, as they have huge latitude in which average they deploy in any given claim. If you want to exaggerate the wealth of this area, you can point at the mean and shout about it. If you want to downplay the wealth, you can rely on the median. If neither of these quite fits the bill, then the mode gives you a handy in-between option.

As it happens, median income is often the most relevant measure, simply because income tends to be highly unequal, with a small number of very large earners, and a much large number of people on a more meagre income. In general, the mean and median will only be the same (or close) if we are looking at a data set with normal distribution, meaning that it is shaped like a bell curve when displayed on a chart. The more the chart is skewed to one end of the range or the other, the more difference there will be between the mean and median.

The mode can also have its uses. Imagine showing a range of people a prototype for a new design of wearable tent (in other words, a coat that converts into a tent). When you ask them what price you think it would be reasonable to charge for the coat-tent, they give you this range of responses:

£50
£40
£35
£30
£25
£25
£20

£20
£20
£20
£20

The median here is £25, but it's possible that you would maximise your sales of the finished product if you could afford to price it at £20 (or even £19.99 if you want to use that old retailing trick to mislead the customers). At that price, you would expect all of these potential customers to see the price as being reasonable, rather than less than half of them.

So mean, median and mode all have their legitimate purposes. But the real point here is that 'average' is a very slippery term that can be used to cover a multitude of evils.

> **Bullet Point: If someone gives you an average figure, check which average they are using or, ideally, ask to see the full data set before you jump to conclusions.**

THE AVERAGE HUMAN HAS FEWER THAN TWO EYES

Just to bang the average drum a bit more, it's always useful to remember this fact – it is true because no one has more than two eyes and some have fewer, so the average person has slightly fewer than two (the mean is probably about 1.9999). This also implies that almost 100 per cent of humans have more than the average number of eyes.

This is another nice example of why you need to be careful about averages, and why different definitions of 'average' can help in different situations: the median or mode number of eyes would almost certainly be two and so would be a more useful version of 'average' than the mean in this case.

Any situation where the distribution is non-standard tends to create confusion over averages. In 2004 the Bush administration boasted that its tax cuts would save the average family $1,586 a year. This was technically true, if you take the 'average' to be the mean. But because the set of earnings across the population included a small number of extremely high figures, 98 per cent of families* actually saved less than $650 as a result of the cuts.

In the same way, in a school that specialises in basketball and thus has a high proportion of tall pupils (but also a smaller number of quite short ones), it might be the case that more than 50 per cent of the children are above the average (mean) height in that school. If you want to know the height that divides the school into two equally sized groups, you would need the median, not the mean.

Bullet Point: The average person doesn't exist and most members of a population set can be above or below average.

WHEN IS 50 PER CENT NOT 50 PER CENT?

Always bear in mind that percentages can be deceptive, especially if they are presented in isolation. In the UK, police numbers fell through the decade from 2010 to 2019 from about 140,000 to about 120,000 (I'm giving rounded numbers to keep this simple). This was a drop of 14.3 per cent (20,000 divided by 140,000). If we are to get the numbers back to the 2010 levels then we would need a 16.6 per cent rise (20,000 divided by 120,000).

So percentage change on the way down is not as big as it is on the way up, because we are measuring the change against a smaller base figure. This can result in very significant differences. For instance, a 50 per cent price cut followed by a 100 per cent price rise will result in no overall change in price from the start to the finish of the process.

The same phenomenon can be misleading in subtly different ways. Imagine three employees at the same company. The CEO earns £500,000, the manager earns £50,000, while the cleaner earns £10,000. At the end of the year the CEO makes a magnanimous announcement. He is awarding a 10 per cent wage rise to the cleaner, a 2 per cent rise to the manager, while restricting himself to a meagre 0.2 per cent increase.

---

* I made up the 98 per cent – the source I was using as a reference for this information merely said 'almost all families' and I used the bogus percentage to dramatise this – see how easy it is to do! A bogus number of this sort can be referred to a Potemkin number – see p. 2.

The Inductive Dodo

A thoughtful sixteenth-century dodo might have looked around its home of Mauritius and drawn some conclusions about the future. The only other animals to be found there were birds, insects and creatures of the sea: the only mammals to be found, due to the remote location of the humanless island, were bats and marine mammals.

The dodo might well have believed that dodos had no natural predators and that the future of the species was secure: if he had been a spreadsheet-minded dodo and had kept population records, he might have projected steady population growth over coming decades. Of course, we know the end of this story: the Dutch colonised the island at the end of the sixteenth century. The arrival of rats, cats and human predators rapidly led to the complete extinction of the flightless bird.

This is an example of the dangers of inductive reasoning. The easiest way to summarise the difference between deduction and induction is that deduction starts from a general principle then moves to a special case, whereas induction starts from a particular instance (or data set) and generates a general principle based on that instance.

The classic example, popularised by Nassim Nicholas Taleb, is the black swan. An English person in the same era as that dodo would have thought it reasonable to assume that all swans were white: it was only in the seventeenth century, when Australia started to be colonised, that it was discovered that there were also black swans to be found in the world. Taleb compares this to the chaos unleashed by the global financial crisis: many analysts and financiers were using models that didn't allow for the supposedly improbable possibility of a sustained collapse in the property market. As a result, they acted with 'irrational exuberance' and the housing crash had huge implications.

Of course, induction will often lead us to reasonable conclusions. I believe the sun will rise tomorrow, because it does every day. I don't know for a fact that it will continue to do so, that the planet might not spin to a sudden halt with my house stranded on the dark side of the planet, but for the most part it is a reasonable assumption. The key thing to remember is that induction can only give us probabilistic ideas about the world, never absolute certainty.

Everyone is cheering his generosity, and drinking to his health after work . . . at least until the cleaner does some sums on the back of a beer-mat and realises that they all got the same £1,000 extra, including the CEO.

On the contrary, if the CEO were to announce that everyone in the company was getting the same 10 per cent increase, their salaries would increase by £50,000, £5,000 and £1,000 respectively. So depending on whether the CEO wanted to be lauded for being a decent person or to maximise his own wage rise, he might choose the way he presents the information carefully.

Bullet Point: Percentages on their own don't tell you very much at all.

THE NON-PROMISE

Do you want to make a promise that is almost impossible to break? Well, one way of keeping a promise would be to make sure you actually do whatever it is you are promising. But that isn't always as easy as it might seem.

So there is an easier method, one that is beloved of retailers and politicians alike . . .

The phrase 'Up to . . .' is incredibly useful when it comes to non-promises. If a shop promises 'Up to 50 per cent reductions' then it has kept its promise so long as at least one item in its entire range is reduced by 50 per cent. (Of course, that might even be an item whose price increased by 100 per cent a week earlier, but we'll come back to retailers' deceptions later in the book.) It's always worth looking closely at any advertisement making this kind of claim: they're technically telling the truth even if they print the 'up to' in a tiny grey font and the 50 per cent in the largest gold shiny font that space will allow.

Politicians can also take advantage of these two little weasel words: if an election manifesto promises to recruit 'up to 20,000 more police officers', to planting 'up to a million more trees' or to increase the minimum wage by 'up to 5 per cent', then you might as well put it in the bathroom to be used as toilet paper because you can be pretty sure those promises are worthless. After all, the politicians will have kept them all if they recruit one police office, plant a single tree in their back garden and increase the minimum wage by tuppence.

And if by chance you've made an actual promise, there are always ways to water it down after the event. In 2019, before the UK general election, the Conservative Party were promising to build an impressive 300,000 new homes a year. When their actual manifesto was published this had been watered down to a promise to build 'at least a million . . . over the next parliament' (which would only be 200,000 a year), while paying lip service to the previous promise by offering to make 'progress towards' 300,000 a year.*

You're making 'progress towards' winning a marathon as soon as you cross the start line, but that is no guarantee that you will be the eventual winner.

Bullet Point: This book will make you up to 100 per cent smarter.

Coronavirus Update

As I write this in late April 2020, the UK health secretary Matt Hancock is on television explaining that the government policy for virus testing is going to be crucial as the pandemic develops. The government got off to a pretty terrible, complacent start in dealing with this issue, in spite of the WHO recommendation that it was crucial. A month ago, we were promised 25,000 tests a day by mid-April. In early April, this was replaced with a grander promise of 100,000 tests a day by the end of April. Of course, the

---

* *Private Eye*, 5 January 2020.

original target was missed. Now we are close to the new deadline, Hancock is assuring us that 'good progress' is being made towards the target and that the government is doing everything it can to meet it. Given that the current daily total is 17,000, one senses that the target is going to be catastrophically missed, but never mind. At least we tried.

Use as Small a Sample as Possible

Did you know that the rates of testicular cancer are highest in small towns?

Did you also know that rates of testicular cancer are unusually low in small towns? Both of these trends have been found in statistical studies, so both most be true, right?

Hmm . . . Clearly something is wrong here. Firstly, let's just note that I was lying in a small way there: the studies have actually shown these trends for 'cancer in general'; I added 'testicular' to make it more dramatic. (I did promise to own up when I was lying, didn't I?)

But that isn't the core problem, which is that studies done in small towns or counties tend to rely on smaller samples. The smaller the sample the less statistical significance the results have and the more likely the result is to diverge markedly from the average result of a much larger group.

Imagine two shops in the same town. One is a supermarket, which has 1,000 customers a day; the other is a corner store, with 200 customers a day. Each records the sex of its customers on a daily basis over a month-long period. I don't know why, perhaps it's something to do with their marketing departments, but just take my word for it that this happens.

Given that the population of potential shoppers is 50 per cent male and 50 per cent female, which of the stores is more likely to have a day on which over 60 per cent of the customers are women?

The data-aware answer is that the small shop is more likely to record such a result simply because the sample is smaller.

Advertisers and pollsters are both fully aware of this tendency. This is why pollsters aim for as large a sample as possible, and give warnings about the likely degree of accuracy of their predictions. Advertisers think about it the other way around. If they have a hundred potential subjects for a piece of market research, they might take the honest approach and use the entire group for a single survey. Alternatively, they could break it down into smaller groups and only report on the group that gives them the most positive response. Or if there are strict regulations forbidding this behaviour in a territory, they might just cross their fingers and try one small group rather than a big one. After all, they aren't obliged to report on any negative results they get, so it is a win-win situation.

People tend to forget to look at the sample size, so are relatively easily fooled by statistically insignificant sample sizes.

Bullet Point: Always ask how big the sample is: if it isn't given, then don't trust any claim based on it.

Rely On Single Numbers in Isolation

A single piece of data rarely means anything in isolation. Imagine you are a keen gardener looking to move to a new city. Would you rely on a report that gave the local average temperature in that city as 60 degrees Fahrenheit and the average rainfall as 750mm?

There is, of course, a lot more you would want to know. There is a huge difference between a city that has about 30mm of rainfall a week and one that has a huge monsoon in August but sparse rainfall for much of the rest of the year. So you'd want to look up the number of rain days a year and the average rainfall by month. Meanwhile, an average temperature of 60 degrees might apply to a very regular, temperate climate, or it might belong to a city that had extremely cold winters and extremely hot summers; so you would want to look at the maximum and minimum temperature and the monthly averages at the very least.

Similarly, if an opinion poll says that 80 per cent of women prefer a particular hair softener, you'd need to know a lot more before taking this as being remotely meaningful. How big was the sample? What is the standard error for this size of sample? What did they prefer the hair softener to? Was it to a rival hair softener, or to mud?

Similarly, if you have problems sleeping and saw a drug being advertised, for which tests showed that on average you would sleep thirty
minutes more per night, you might be tempted to give it a try. But if it
turned out that this was a survey of 100 people, seventy-five of whom
slept an extra hour a night, and twenty-five of whom slept an hour less
per night and that the effect persisted after you stopped taking the medication, you might pause for thought as to whether odds of 1 in 4 for a
worse sleeping experience were worth the risk.

**Bullet Point: Ask for as much detail on a data set as possible, and
never take a single number at face value.**

DON'T REVEAL WHAT QUESTION YOU ASKED

This isn't really a question about numbers, but it's an important thing to
be aware of when considering the results of a survey or poll. Consider
the following two ways of asking a question:

Do you want free healthcare provided by the state if you get sick
or injured?
Do you want to be taxed an extra 5 per cent a year to fund univer-
sal healthcare?

This is a fairly extreme example, but the way the question has been asked
could clearly affect the responses. In general, this is a problem that pollsters struggle with, as responses to a particular question can be affected
by many different aspects of the way it is framed. For instance, a poll in
which the first few questions are about crime and punishment may lead
to a stronger performance by the party most associated with law and
order in the next few questions.

There are subtler instances: one advertising case involved Colgate,
who in 2007 were ordered by the British Advertising Standards Authority
(ASA) to withdraw a billboard ad that claimed that 'More than 80 per cent
of dentists recommend Colgate'. The ASA pointed out that the claim
'. . . would be understood by readers to mean that 80 per cent of dentists

recommend Colgate over and above other brands, and the remaining 20 per cent would recommend different brands'.

Which sounds about right, doesn't it? But what had actually happened was that the survey asked dentists to name multiple brands, so it was perfectly possible that a similar number had recommended other brands in the same survey. The deception may or may not have been deliberate but it was certainly a misleading claim to make, thus the slap on the wrist from the ASA.

In a similar vein, in the 2019 UK General Election, the Liberal Democrat Party published a poll that appeared to show that they were so far ahead of the Labour Party in the potentially marginal North East Somerset seat that anyone wanting to defeat the incumbent Conservative candidate (the widely hated Jacob Rees-Mogg) should vote for them. What the small print showed was that the question asked had actually been a blatantly leading one: 'Survation polled 405 respondents aged 18+ living in NE Somerset with the question: "Imagine that the result in your constituency was expected to be very close between the Conservative and Liberal Democrat candidate, and none of the other parties were competitive. In this scenario, which party would you vote for?"' When the party leader Jo Swinson was challenged on a live news programme as to how she could criticise misleading claims such as the '£350 million a week' Brexit promise (see p. 57) while endorsing such blatantly misleading statistics, her weak answer was, 'It's different.' (Her generally weak performance in the campaign led not only to a poor Liberal Democrat performance in the election, but to her losing her own seat in the process.)

Bullet Point: You need to know the questions, not just the answers.

Avoid Using a Common Baseline

One of the reasons why statistics in isolation are meaningless is that, unless we already have a good understanding of the subject in question, we need something to compare that statistic to: a common baseline. For instance, there are 140,000 red squirrels in the UK. Does that sound like a lot or not very many? Well, it would depend what we are comparing it to. If we add that there are currently 2.5 million grey squirrels (which were introduced in the 1800s and which tend to drive out red squirrels from an area) and that the red squirrel population is thought to have peaked at 3.5 million during the twentieth century then we can see that the current population is a pretty low number, which indicates a species that has declined significantly. For more detail on the sustainability and direction of the current population we'd need to see the numbers for the last few decades compared to each other.

Similarly, consider the measurement of GDP. Putting aside the complexities of the way this is measured by economists, which is a can of worms in the first place, suppose we were told that in 2018 the GDP of the UK was £2,179 billion, while the GDP of Sweden was 5,289 billion krona.

This basically tells us nothing, because it is an example of comparing apples with oranges. So first, let's convert both to dollars and we get $2,828 billion versus $556 billion. So the economy of the UK is bigger, which means the UK is a wealthier country, right?

Well, it's best at this point to reduce this further to a common baseline, which is GDP per capita (meaning per person). Then we get Sweden with $54,346 per capita as opposed to the UK with $42,580 (at the exchange rates available at the time of writing). So while this isn't a perfect measure of wealth, and it is always worth looking at as many alternative sources as possible, the suggestion at this point, that Sweden is wealthier than the UK, is at least based on a valid comparison.

Here's another way things can go awry. In the official statistics published by the Office for National Statistics (ONS) in the UK, 2017 saw a seemingly significant fall in the number of alcohol-specific deaths from the previous year. For males these fell by 18.7 per cent, while for females it was a whopping 24.6 per cent fall. So what did this reflect? A major reduction in drinking? A successful change in the healthcare system?

No. In fact, it reflected something far more mundane: the definition of 'alcohol-specific deaths' had been amended in the meantime so that it excluded a range of causes of death that were only 'partially attributable to alcohol'. So in this case we can disregard the fall from year to year.

So bear in mind that for a common baseline to be useful we also need to be sure that the way the particular statistic is being measured is consistent. In this case, it is worth mentioning that the ONS was making the change for an entirely valid reason: they wanted to harmonise the definition being used around the UK with the standard measures used in other countries, which in the end meant they were using a statistic that was more useful and comparable. But it would be easy for the casual user of their statistics to be misled if they didn't check the small print.

Bullet Point: If you want to befuddle the consumer about a statistic, the best way to do this is to avoid giving common baselines or clear comparisons at all costs.

CORONAVIRUS UPDATE

The use of a standard baseline is not always enough to make a set of data useful or informative. During the pandemic we have regularly been shown a graph comparing the overall deaths from Covid-19 in different countries around the world. The experts have been scrupulous in that the lines on the graph all start at the point where there were fifty deaths in a country. And you can gain some insight into the pandemic from this, for instance it is remarkably clear that countries like Germany or South Korea, both of which took the WHO advice to prioritise mass testing, have succeeded in keeping the death rate far lower than countries like the UK or the USA, both of which had confused responses bordering on denial from the centre of government in the early stages. But the graph glosses over other key issues: for instance, the number of deaths per million of the population is likely to tell us a lot more about the effects of government policy than overall death rates. And the ways in which the count of deaths is compiled in different countries differs to the point that the graph can be misleading: Belgium currently counts all deaths in which the virus is likely to have played a part, while the UK continues to focus on deaths in hospital only. The experts, of course, will have access to different permutations of data. But, as ever, one can wonder at how the simple range of graphs presented for public consumption was chosen and why.

Use a Biased Sample

We saw earlier how that famous Camel advert was based on research using a sample of doctors who had just been given a free box of Camels. That is one obvious way to use a skewed sample to distort reality. But there are many other ways of using a sample that is likely to be biased.

For instance, if you are looking for a sample that is biased when it comes to the question of social care, then you might want to use a sample that is heavily skewed towards retired people. And if you want a sample that would be inclined to support the legalisation of marijuana, you might want to use a group of college students, ideally from one of the more liberal colleges or universities.

Bias can, of course, be intentionally introduced. But there are many subtle ways in which bias can creep into a sample. If you only use landline phone polling, you are probably going to reach a disproportionate number of older respondents, since many younger people rely on mobiles. If you stop people on a random city street (an easy, quick way of finding a sample) you will immediately be introducing a bias towards the kinds of attitudes that urban dwellers have, but there might be all kinds of other unforeseen biases: for instance, you might be polling right next to a big computer company, a political party headquarters, an arts college or a homeless shelter. Each of these would introduce different kinds of bias.

Polling companies go to great lengths to reduce the impact of bias on their samples, profiling respondents and using that data to attempt to weight the responses of different people so that no group is overrepresented in the sample. Of course, this isn't a perfect process: people are too complex for any profiling system to fully rule out bias, but it can produce a better result that an unweighted one.

But the polling companies (mostly) want to get at the truth rather than justify lies, while there are others such as advertisers and politicians whose motives are less pure. So if a statistic is being used to support a dubious claim, it is always worth investigating how the sample was gathered as well as checking what questions were asked, and to ponder whether or not the sample might have been biased.

Bullet Point: If you want a sample that supports the abolition of Christmas, you might want to do your polling on a turkey farm.

Exploit Cognitive Biases

There are numerous psychological effects that can be exploited when you are trying to mislead someone. In a classic psychology experiment, originally performed by Solomon Asch in the USA, a group of people were asked to assess the length of lines printed on two cards. While standing in a semicircle with other participants, they would be shown two cards. One would have three lines of varying lengths: for instance, they might be 8 inches, 10 inches and 12 inches long. Then they would be shown a second card with, for instance, a 10-inch line on it. Then the participants would be asked to choose the line that matched the second card from the first card.

The catch was that there was only one real participant in this study; the others were actors who had been instructed to choose the wrong line: for instance, they might all indicate the 12-inch line was the one that matched the second card. The real subject would thus be placed under a peculiar type of peer pressure: people find it difficult to be the only one of a group to have a different opinion. In the original experiment only a quarter of the real subjects chose the correct line on each of twelve trials. And as many as three-quarters of them chose the wrong one at least once.

This tendency to **groupthink** is the motivation for those advertisements or political broadcasts that use a succession of 'normal' people all singing the praises of a product or candidate. We have a natural tendency to want to agree with the crowd, especially if the actors representing the crowd happen to be well enough chosen to make us recognise them as being 'people like me'.

The effects of in-groups and out-groups on our thinking also can't be overestimated. I mentioned earlier the claim during the Brexit campaign that we could take back £350 million a week from the EU and give it to the National Health Service. In spite of it being debunked extensively, the claim was still believed later by 42 per cent of people who had heard it. There were probably two effects at work here: firstly, people's attachment to their political beliefs can outweigh their ability to calmly analyse facts, as we noted earlier in the section on motivated numeracy (see p. 6). And secondly, we suffer from **confirmation bias** (which happens when we look for evidence that our existing beliefs are correct rather than looking for counterevidence) and **cognitive dissonance** (which makes it uncomfortable for us to recognise when one of our existing beliefs is wrong).

There are numerous other cognitive biases that can be used to pressure someone into believing a lie or making a bad decision. For instance, the **primacy bias** refers to the way that the first piece of information we encounter sets the tone for how we interpret the rest of the information we are presented with. Imagine, for instance, that I describe one of my fellow maths authors* thus: arrogant, stubborn, irascible, precise, intelligent, deep-thinking.

Now imagine that I describe the same person in these terms: deep-thinking, intelligent, precise, irascible, stubborn, arrogant.

You may well have got a better impression from the second list than from the first, even though the lists are essentially the same (and psychological studies have confirmed this tendency). The first term in the list tends to set the tone, and the rest of the information is interpreted through this prism. So the first author sounds like an arrogant boor whose intelligence I grudgingly admire, while the second sounds like something of a genius whose flaws are part of the package.

We saw earlier how the weasel words 'up to 60 per cent' can be used to make a sale in a shop seem more generous than it really is. Retail

---

* You may or may not be burning with curiosity as to who I am slagging off this way. Honestly, this was a deliberate bit of misdirection: I may be thinking of a particular person, but I know the lawyer would pick me up on it if I gave any information that could actually identify them, so I pretended to be talking about a fellow author. Lying is fun, isn't it?

outlets can use a similar trick when it comes to the layout of the shop. My local supermarket positions a long shelf of special offers right inside the door: you are funnelled past these to reach the aisles. Each aisle then has at least one special offer at the end of the aisle, facing into the main route through the store. In each case, we encounter the most reduced items in the store on our way to the less generously priced products we actually need to buy.

The **recency bias**, by contrast, is our tendency to remember the most recent piece of information we are presented with. This may seem like a contradiction, but if you bear in mind that we are most likely to remember the first and last items on a shopping list and to be forgetful about the items in the middle, then you can make sense of how both of these biases can apply.

One way this can be exploited is in the way we order the facts we are presenting. Personally, I hate PowerPoint presentations, largely because they seem designed to be simplifying and misleading: the kind of chart or list that works in a single slide tends to be the most basic, reductionist summary of a situation, and therefore the one most likely to avoid drawing your attention to any problems with the way the statistic is being deployed.

Anyhow, sales managers are very fond of them, so let's imagine a sales manager who has to present the following information:

1. Number of items selling 10,000 units: 17 (as opposed to 21) in the previous year
2. Year-on-year growth in profit: -2 per cent
3. Year-on-year growth in volume: 6 per cent
4. Year-on-year growth in turnover: 4 per cent
5. Growth in exports volume: 10 per cent
6. Growth in domestic sales volume: 1 per cent

It's clear that this is a pretty mixed bag: some good news and some bad news. But the sales manager might make the best possible fist of it by tweaking the order thus:

1. Growth in exports volume: 10 per cent
2. Growth in domestic sales volume: 1 per cent
3. Number of items selling 10,000 units: 17 (as opposed to 21) in the previous year
4. Year-on-year growth in profit: -2 per cent
5. Year-on-year growth in turnover: 4 per cent
6. Year-on-year growth in volume: 6 per cent

This way the sales manager starts with some upbeat news, and can congratulate the exports team, creating a feel-good start: this softens the blow of the two weaker items to follow. The profit news is obviously bad, so gets buried just before the final double whammy of better news, allowing the sales manager to subtly imply that things are on the up and increased profits are bound to follow shortly. And that final graphic can be left up on the screen while the sales manager gives a pep talk, so the recency effect is reinforced.

Next we have to consider the **availability bias.** This concerns the way that we store and recall information. The most easily available data is given undue weight in our assessment of a situation. For instance, in areas prone to earthquakes such as California, the number of people taking out insurance against earthquake damage rises sharply shortly after an earthquake, then steadily falls as the most recent instance of an actual earthquake recedes in people's memories. This is not rational behaviour, but it is understandable as people are acting on the most available information.

Think also of how we rate the danger of our daily activities. You might consider riding on a rollercoaster or theme-park ride to be fairly dangerous: those incidents that do occur receive a lot of media coverage and are nasty enough to stick in the memory.

However, it is almost certainly not as dangerous an activity as you might assume. The American organisation, the International Association of Amusement Parks and Attractions, who would of course have a vested interest, carried out what at least appears to be fairly reliable research, which recorded eight injuries per million activity days.

As a single statistic this means very little, of course. But they provided a comparison point with playing American football (343 injuries per million activity days); it even turned out that fishing was a more danger-ous activity, with 88 injuries per million activity days.

There's a connection here with another cognitive bias: the **bizarreness effect**. This essentially means that the more garish a particular thing is the easier we find it to remember. Being bitten by a snake, being shot or assaulted by a stranger, having your passenger plane shot down by a rogue missile, being attacked by a shark . . . These are all the kinds of strange, unusual events that do really happen from time to time, and which many people are fearful of.

But far more people die (on pretty much any measure) driving in a car, climbing the stairs or cycling to work. We are far more likely to be shot or assaulted by someone we know than someone we don't know. Smoking a cigarette or consuming excessive amounts of alcohol hugely increases our chances of dying of lung cancer, cirrhosis of the liver, stroke or heart attack in the future, but this doesn't scare us as much as it should, which is why so many people who might be terrified of being attacked by a crocodile will smoke and sip from a beer while walking fearfully alongside a bayou.

There is a mathematical measure of a one-in-a-million chance of death: the 'micromort'. Here for comparison are the micromort statistics for a few ways of dying (these are from the UK, but similar figures apply in many countries, including America):

Your chance of dying from natural causes on a given day: 24 micromorts

Your chance of dying from any non-natural cause on a given day: 0.8 micromorts

Your chance of being unlawfully killed on any given day: 0.027 micromorts

When it comes to transport, we can compare journey lengths on the basis of micromorts. Each of the following increases your chances of death by one micromort:

6 miles on a motorbike
17 miles as a pedestrian

Approximately 15 miles on a bicycle
230 miles by car
1,000 miles by airplane
6,000 miles by train

So clearly, some of the things we are scared of are far less probable than the things we forget to be scared of. Thoughts of everyday, normal activities like driving or smoking scare us less than more readily available thoughts of causes of death.

There is another aspect to this: when it comes to illegal drugs, the media has a tendency to play up the idea that all drugs are equally terrifying. Over recent decades they have focused on deaths among ecstasy and cocaine users, perhaps because some of the people who have died using those drugs have been more glamorous, and higher status. The figures below come from a combination of 2009 NHS and ONS figures (with a few assumptions on questions such as how many users there are of particular drugs). They show the micromorts per monthly user of a drug per year:

Ecstasy: 224
LSD: 0
Heroin: 5,616
Cocaine: 109
Cannabis: 0.65
Methadone: 4,480

One well-known British drugs expert (David Nutt) received a lot of criticism when he observed that taking ecstasy was only about as dangerous as riding a horse. This didn't fit with the media tendency to demonise drugs users, and he was described as trivialising the dangers of drug-taking, but his comments were simply based on the statistics. It could be argued that government policy and media attitudes should likewise be based on the statistics, but it's probably best not to bet on that happening any time soon.

In the meantime, we will probably all continue to be scared of the wrong things and to keep on smoking, drinking, eating too much, driving dangerously and generally behaving recklessly with our own personal safety.

Bullet Point: If you do die today, it probably won't be due to a snake bite or a stranger attack.

Coronavirus Update

The **anchoring** effect is in play when our perceptions of a number are affected by the initial numbers mentioned. For instance, if you ask a set of subjects in an experiment how much they would pay for a piece of jewellery, but rig the group so that the first subject gives an unreasonably high estimate, this will lead to the other subjects also overestimating the value. At one point in the pandemic we had UK government ministers suggesting that a death rate of 20,000 would be a 'success' on the basis that this is about how many people die in an influenza outbreak. This wasn't a successful piece of news management since we are currently on target to tragically overshoot that figure considerably. But possibly the most extraordinary moment so far came in the press conference in which President Trump predicted 100,000 to 200,000 deaths, but argued that without his interventions it would have been closer to 700,000. So that's a success, then . . . The fact that most scientific experts consider that the actions of his government were complacent to the point of allowing the outbreak to get out of control in the first place doesn't feature heavily on Fox News, so it probably hasn't tickled the president's conscience at this stage.

Looking for Counterexamples

It's worth taking a while to explore the effects of confirmation bias as it is particularly useful for understanding the kind of irrationality that can lead us to miss obvious lies. Firstly, imagine you are given this sequence of numbers:

12, 14, 16

You are told it is an example of number sequences that obey a particular rule and you have to guess what that rule is. You have to come up with some number sequences and you will be told whether or not they conform to the rule. At any stage you can guess what the rule is and will be told if you are correct. What sorts of number sequences do you think you would choose to try to find out what the rule is? Before reading on, it's worth writing down two or three examples.

Right, so there has been a lot of experimental research on this. What most people tend to do is to first come up with another sequence that is obviously similar, such as a sequence of three even numbers in a row, like:

2, 4, 6

or

18, 20, 22

Others may be bold enough to try some odd sequences, also ascending by two each time. A few use other strategies, but not many.

So, after testing out a few of these sequences, people's confidence that they know the rule tends to rise: they will guess that the rule is 'Three even numbers rising by two', or 'Any three numbers rising by two', only to be told this is wrong.

The really interesting thing is what comes next. Some people start to experiment with other kinds of sequences separated by two, like 13, 11, 9. Some will try out other theories about numbers rising by the same amount each time. Some will stick with a rule that is based around even numbers. Some will even attempt to restate their original theory in a novel way, like 'three numbers falling by two at each step from the last to the first'.

Eventually, the experimenter will reveal the rule. In this case it is 'Any three numbers in numeric order', so '1, 4, 2007' or '33, 87, 541' would equally have fit the rule. (And of course there are many other rules that this sequence could have fit, like 'Any three numbers', 'Any set of numbers in numeric order', and so on.)

The point is that when faced with this problem, people start with an assumption and then try to **prove** it, where they would get a lot further a lot quicker if they attempted to **disprove** it. There is no number of examples of the rule 'Even numbers rising by two' that would prove the theory correct, but it only takes one counterexample to disprove it.

Just to dig a bit deeper, let's look at another experiment which, like the one above (and several others mentioned in this book), was recorded in Stuart Sutherland's brilliant book *Irrationality*. This one was originally performed by Peter Watson. He would put four cards on the table, reading:

A B 3 4

The subject would be told that every card had a number on one side and a letter on the other. They would then be asked which two cards they would turn over to establish the truth of the statement: 'Cards with an A on one side have a 3 on the reverse side.' Most people pick the cards that read A and 3.

They are correct to pick the A: if it has a 3 on the other side, it would be some evidence for the theory, and if it doesn't it would be a counterexample. However, the 3 is no possible use in checking the theory because even if it has another letter on the reverse, it only proves that there are other letters that sometimes have 3s on the back. There is also no point in them picking the B, since whatever number it has on the reverse won't be of any relevance. However, the thing that most people miss is that the second card they should pick is the 4: if this one has an A on the back, then it would be a counterexample.

(It's worth mentioning that the availability bias may also be at work here. A later psychologist tested this by using the same set-up, but asking subjects to test the rule: 'Cards with an A on the back don't have a 3 on the reverse side.' In this case, the A and the 3 are logical choices, and there was still a tendency for subjects to choose those cards. But the evidence for confirmation bias in our thinking comes from so many different experimental sources that this is only a minor quibble.)

Our tendency to look for confirmatory evidence rather than counterexamples is deeply ingrained, and may be a large part of the reason why we can fall subject to motivated thinking. For instance, when we strongly hold a belief, we even view evidence for and against it differently. Presented with two equally flawed studies on an emotive topic such as

Brexit or gun control, we will notice all the flaws in the study that goes against our instincts, while failing to see the flaws in the one that supports our side. Here's another maths-based example of this.

In one experiment, subjects were shown two bags. They were told that one had 60 per cent of black balls and 40 per cent of red, while the other had 60 per cent of red balls and 40 per cent of black. They were given one of the bags and had to take balls out one by one. At the point when they formed a hypothesis as to which bag this was, they were asked to tell the experimenter their hunch. Then they would draw two more balls. The focus of the experiment was on what happened when the next two balls came out as one red ball and one black ball.

Now, this is obviously entirely neutral evidence. As a random sequence of two balls it would be just as likely to happen whichever bag you had in front of you. But when the subjects were now asked whether they were more certain, less certain or neither more nor less certain of their initial hypothesis, almost all of them reported that they were more certain.

What this shows is that once we have come up with a hypothesis, we are capable of interpreting neutral evidence completely differently to how we would if we held the opposite belief.

Bullet Point: Always look for evidence that would disprove your initial assumptions.

THE TWO-BY-TWO TABLE

There is a wider problem that is created by our blindspot for negative evidence. Let's look at the possible associations between two events: to avoid calling the events or properties X and Y, let's say they are 'Having brown eyes' and 'Being right-handed'. The four possible combinations of these properties in a person are:

Has both right-handedness and brown eyes.
Has right-handedness, doesn't have brown eyes.
Has brown eyes, doesn't have right-handedness.
Has neither right-handedness nor brown eyes.

Not Falsifiable

On the subject of confirmation, it's worth remembering Karl Popper's view that a scientific idea is essentially meaningless if it is not falsifiable: 'The discovery of instances which confirm a theory means very little if we have not tried, and failed, to discover refutations. For if we are uncritical we shall always find what we want: we shall look for, and find, confirmation, and we shall look away from, and not see, whatever might be dangerous to our pet theories. In this way, it is only too easy to obtain what appears to be overwhelming evidence in favour of a theory which, if approached critically, would have been refuted.' Carl Sagan gives the example of a man claiming to have an invisible dragon in his garage: it is invisible, floats in the air, so doesn't leave footprints, can't be felt and breathes invisible, heatless fire. How could you prove it isn't there?

Let's imagine an alien detective is observing a group of 200 people who fall into all of these categories. We can lay the information about them out in a two-by-two table thus:

	Brown eyes	Not brown eyes
Right-handed	120	30
Not right-handed	40	10

The alien keeps meticulous notes as he is trying to understand the human species. At the point when he has noted down 150 right-handed people of whom 120 have brown eyes, he concludes that he has good evidence that there is a reasonably good, if not perfect, correlation between right-handedness and brown hair and suggests an experiment involving entirely unnecessary vivisection to test out his theory.

The alien statistician yawns and tells him that since he has only investigated cases in which people are right-handed, he hasn't shown any particular correlation. He needs also to investigate people who aren't right-handed.

So the alien goes back and finds 50 people who aren't right-handed of which 40 have brown eyes. He reasons that since he has found 160 people with brown eyes, of whom 120 are right-handed, he still has some pretty strong evidence for the correlation, and it's definitely time to get the vivisection table ready.

The statistician, who also happens to be his boss, takes him off the case and puts him on sentry duty until he can learn some basic statistics. The numbers he has brought back, as can be seen in the table above, merely show that the proportion of people with brown eyes is exactly the same among right-handed and left-handed people. So there is zero correlation between the two properties.

This is a cognitive bias that can have serious consequences. In one experiment nurses were asked to look at 100 cards on which it was recorded whether a patient had a particular symptom and a particular disease.

The two-by-two table was:

	Patients With Disease	Patients Without Disease
Patients With Symptom	37	33
Patients Without Symptom	17	13

In the study, 85 per cent of nurses thought that the symptom was an indicator that the disease was present. Presumably they were swayed by the fact that the largest number of cards showed patients with both the symptom and the disease. But a clearer analysis of the data shows that there is a similar proportion of patients with the symptom among patients with and without the disease, which actually indicates no connection.

It's worth noticing that journalists and advertisers are extremely prone to making outrageous statistical claims that would be easily debunked using a two-by-two table, or simply by considering all the possibilities. Consider, for instance, a news item that claims that 'Travelling in London's Rush Hour Increases Chance of Heart Attacks'. Let's say that this is based on a statistic that reveals that 20 per cent of all heart attacks on London transport happen during the rush hour.

This may or may not be significant. The only way to find out would be if we know what percentage of all journeys on London transport are made during the rush hour. If it is around 20 per cent then we can safely assume the journalist is talking nonsense.

Bullet Point: When considering if two events are related, make sure to take all four possibilities about their connection into account.

CONDITIONAL PROBABILITY

Two-by-two tables can be extremely useful when it comes to understanding the difficult concept of **conditional probability.** For instance, let's say we have the following information. At a pet store, 100 customers make a purchase. Out of those hundred, 40 bought cat food, 30 bought dog food and 10 bought both cat food and dog food. The question is, if a customer bought cat food, what is the probability that they also bought dog food. Here's the two-by-two table:

	Bought cat food	Didn't buy cat food
Bought dog food	10	20
Didn't buy dog food	30	40

By laying the information out this way, we can fairly easily see that among customers who bought cat food (40 in total), 25 per cent (10 customers) also bought dog food. We would express the probability as 0.25.

The formula for conditional probability is:

$P(B|A) = P(A \text{ and } B) / P(A)$
which can also be expressed as:
$P(B|A) = P(A \cap B) / P(A)$

In this equation A∩B or P(A and B) means the probability of A and B both being true, or in this case the customer having bought dog food and cat food. P(A) means the probability of A being true, while P(B|A) means the probability of B being true if we know that A is true.

So P (customer buying dog food, given that they bought cat food) = (0.1)/(0.4) = 0.25.

One example of faulty reasoning comes when we have to consider two variables in a situation. Imagine a case where a bystander has witnessed a hit-and-run accident, in poor lighting conditions. As it happens, the town has only two colours of car, of which 85 per cent are blue and 15 per cent are red.

She thinks that she saw a red car, but for safety the investigating team set up a test, and find that the bystander only correctly identifies the colour of a car 80 per cent of the time (and 20 per cent of the time she incorrectly thinks a blue car is a red car). Is it more likely that the car was red or blue? Based on the bystander's testimony, most people tend to answer that it is more likely to have been red. But let's run through the conditional probabilities, focusing on the two outcomes in which she thinks she sees a red car.

The chances that the car is blue are 85 per cent, while the chances that she has seen a blue car and identified it as being red are 20 per cent. In this case we multiply the probabilities together and find that there is a 17 per cent chance of her having seen a blue car and identified it as red.

The chances that the car is red are 15 per cent, and she would correctly identify this 80 per cent of the time. We multiply these together and get 12 per cent. So the probability of the car being blue is actually 0.17 as opposed to the 0.12 chance of it being red.

Again we can lay this out in a two-by-two table, using 100 cars as our base figure:

	Identified as blue car	Identified as red car
Blue car	68	17
Red car	3	12

Note that the conditional probability of the bystander having seen a red car (given that she claimed it was a red car) is calculated using the formula above as 12/29 which is about 0.43 as opposed to the 17/29 = 0.57 that it was a blue car.

The classic case in which conditional probability is hugely important is when it comes to understanding tests for particular diseases. There are tests for some diseases such as Huntingdon's Disease, which can develop later in life. Imagine a disease called Bayes' Disease (named after the great mathematician who developed conditional probability theory). We test 10,000 patients for a particular gene XYZ, which can indicate that Bayes' Disease will develop later in life.

	Have XYZ gene	Don't have XYZ gene
Will develop Bayes' Disease	99	1
Won't develop Bayes' Disease	20	9,880

Now the first thing to note here is that almost everyone who is going to develop the disease has the XYZ gene. There is thus a 99 per cent possibility that if someone is going to get the disease, then they have the gene. The big danger here is that most people (including many doctors in tests that have been run to simulate similar problems) are prone to making the inference the other way around: they assume that if you test positive for the XYZ gene then you are 99 per cent likely to develop the disease.

In fact, if we use the conditional probability calculation, we have 119 people who have the XYZ gene, 20 of whom won't develop the disease. This means that if you test positive for the disease, then you only have a 20/119 = approximately 0.17 chance. So firstly, it is hugely important for doctors to truly understand the maths behind both false positives and false negatives when it comes to testing for any kind of medical problem, as this will be crucial when it comes to giving the patient a clear picture of the situation. It is also possible for this information to be made clear before testing: some patients may choose not to take a test if a false positive will leave them in such uncertainty and if there are no further tests that can confirm the situation. And where further tests are available, the patient's mental welfare will be significantly affected if the situation is not clearly explained.

Bullet Point: P(B|A) = 100 per cent where B is 'not understanding conditional probability' and A is 'making or believing egregious numerical mistakes or lies'.

That Random Feeling

People are generally pretty poor judges of probability and randomness – as a species we have largely evolved by learning to recognise patterns, and this leads us to imagine patterns that aren't there when presented with something genuinely random.

In addition, we are bad at intentionally creating a 'random' series. Think for a moment about which of these sequences of results of coin tosses is the most likely: HHHHHH, TTTTTT or HTHHTT. Most people instinctively see the last as the most probable, failing to realise that each is equally likely. (Since the probability of heads and tails on each throw is ½, any of these sequences would occur about once in every sixty-four times you make six consecutive coin tosses.)

This perception also feeds the **gamblers' fallacy** in which, when asked to predict what the next coin toss will be after the series HHHHHH we tend to overestimate the chances of tails, because of our innate feeling that the results will somehow 'even up'.

In the 1890s, the Monte Carlo roulette results used to be published daily in the newspaper *Le Monaco*. The mathematician Karl Pearson wanted to test some of his methods on random data, so used the roulette results he found in the newspaper. He found that the results were so peculiar that they could only have resulted from a heavily rigged or biased wheel. In fact, the lazy journalists at the newspaper had decided that no one would notice if they simply made up the results, so rather than uncovering dark deeds at the casino, Pearson had merely demonstrated how bad the journalists were at making up random data sets.

When we try to imagine how a random process will look we tend to come up with a series of results that are consistently erratic, that don't contain anything that 'looks like a pattern' and that spread the results around fairly evenly. In fact, a genuine random series of roulette wheel spins can look quite clumpy, with series of results that look very 'pattern-like'. For instance, the ball might not land on the number 7 for hundreds of spins, in spite of having landed frequently on the number 6. Genuinely random events often don't look random to us.

A related problem afflicted Apple when it introduced the shuffle feature on the iPod. Many users noticed songs by the same band occurring close together, or the same song recurring, and complained, wrongly assuming that the shuffle was somehow rigged rather than being truly random. Steve Jobs was offended by the mathematical slur, but nonetheless amended the product, explaining that they had had to make the shuffle 'less random to make it feel more random'.

Bullet Point: Bear in mind that random data can look like it contains patterns and vice versa.

PROBABILITY SCAMS

While we're on the subject, it's worth bearing in mind that many common scams are based on our lack of an intuitive grasp of probability. The betting prediction scam, for instance, involves the con man sending out, say, 32,000 emails to gamblers predicting the result of a tennis or boxing match, with half of them predicting one outcome and half the opposite. Then another 16,000 are sent out, again predicting a match with a 50 per cent divide between the two outcomes. This continues until there are just 1,000 recipients of the emails, who at this stage are meant to be highly impressed by the con man's brilliant run of predictions: the con man can now offer to continue sending their 'tips' for a considerable charge. If enough gamblers fall for the con, there is a large profit to be made.

Scammers who use dice or cards can use similar probability-based methods. Dice odds can be confusing. For instance, you might think the odds of throwing a 12 with two dice are identical odds of throwing an 11. Each is only feasible with one combination of numbers, (6, 6) and (6, 5). But taking into account the probability on each dice, you can see there is only one way to throw a 12 (6, 6) and there are two ways to throw an 11 (5, 6 and 6, 5). So if you are offered an even-money bet that you can throw a 12 before the other player throws an 11, it is a pretty bad bet: your odds are 1/36 and the other player's is 1/18 (as there are 36 possible combinations of dice).

Maverick Solitaire, named after the television show *Maverick*, involves picking twenty-five random cards from a deck and trying to arrange them into five sets of five, each of which is a straight or higher at poker. This sounds like a long shot, but it is actually 98 per cent probable that you will be able to do it: there will always be at least two flushes you can find immediately, and it will usually be possible to find straights and full houses without too much difficulty. Again, best not to bet against someone doing this.

Another card hustle is the royal bet – you cut a deck of cards into three piles, and bet that at least one of the cards will be a jack, king or queen. There are only twelve of these in a pack of fifty-two, so it sounds like it might be less than 50 per cent likely. But you have a 12/52 chance for the first card, a 12/51 chance for the second if that fails, and 12/50 on the third if need be. These are cumulative odds, which can be added together to show that the odds on finding at least one royal are approximately 70 per cent, so again the odds are counter-intuitive, unless you have an instinctive understanding of probability.

Bullet Point: If someone offers you a bet that sounds like a winner, work out the odds carefully.

APPEAL TO AUTHORITY

There are a few cognitive biases that mean that we tend to judge a message based on who is delivering it. Broadly speaking, the **halo effect** is the tendency for a person's positive or negative personality traits to affect how we perceive other, unrelated aspects of the personality or performance. And the physical attractiveness principle is the tendency to assume that if someone is physically attractive, they possess other, unrelated socially attractive traits.

These are all exploited to varying degrees by advertisers and television shows. For instance, the most attractive participants in reality television shows are often called into magazine shows as experts on dating or emotional problems; attractive performers are used in adverts to convey an aura on whatever product or service they might be selling; and scientific experts are used to bolster the claims being made about a product.

In some cases a combination of these effects is being exploited. That attractive dentist assuring us that 70 per cent of their patients say that a particular toothpaste reduces the sensation of burning gums may simply be an actor portraying a dentist, but our unconscious biases still lead us to accept his word more readily than we might from an anonymous voice-over. A real scientist presenting the pseudoscience that often underpins beauty products is likely to convince us that the science is legitimate.

Just as an aside, a recent news item that isn't anything to do with maths gave a beautiful demonstration of the danger of trusting in authority. During the dreadful bushfires of 2019 in New South Wales, the Scottish ITV News Asia reporter Debi Edward was persuaded by the staff of Kangaroo Island Wildlife Park to wear full-body protective gear so she could safely do an item to camera holding a 'drop bear', a vicious bear related to the koala that has vicious claws and a poisonous bite. Clearly nervous, Edward completed the report to camera before being alerted by the laughter of the crew that she had been taken in by a widespread Australian hoax: tourists are often warned of the vicious drop bear that falls on unsuspecting passers-by from trees, and told they can ward it off by, for instance, smearing Vegemite behind their ears, wearing forks in their hair, or urinating on themselves. All Edward had actually been holding was a harmless koala.

And bear in mind that when celebrities appear on camera they may have been taken in by the 'authorities' who have told them what to say. In the notorious 'Paedogeddon' episode of the brilliant UK TV comedy *Brass Eye*, the UK DJ and TV presenter Neil Fox was one of several celebrities who were fooled by creator Chris Morris into reading out ludicrous statements that skewered the contemporary media scares about paedophiles and the willingness of celebrities to involve themselves in issues they simply didn't understand. So Fox found himself solemnly reading out the line: 'Paedophiles have more genes in common with crabs than they do with you and me. Now that's a scientific fact: there's no real evidence for it, but it *is* scientific fact.' If you happen to be the celebrity fronting a campaign of any sort, make sure you at least take a moment to understand what message you are actually sending and that it isn't mathematically or scientifically preposterous?

At a more extreme level, celebrities can also be used illegitimately in 'fake news' to endorse dodgy products. At the time of writing, both Daniel Radcliffe and Kate Winslet have had to distance themselves from fake online adverts for dubious cryptocurrency tools that claim they can make you a Bitcoin millionaire with no expertise or knowledge. In each case, a fake television interview is referenced in which the actors were claimed to have endorsed the products being sold.

There's one obvious moral in these particular instances: if a product (especially a financial product) sounds too good to be true, then it almost certainly is. I'll return to the whole subject of financial scams such as pyramid schemes later in the book, but almost all of them rely on the idea that you can get rich without hard work or experience. In this case the size of the numbers being bandied about can lead people to suspend their natural caution and jump into a bad decision, and the celebrities being falsely used to sell that message are inadvertently lending their halo effect to misleading claims.

> **Bullet Point: Imagine whether or not you would believe the message if it was delivered by a random unattractive stranger rather than an expert or celebrity.**

Conclusion: Take a 360-degree Look Around

In this chapter we've taken a tour through some of the basics of how statistics and numbers can be used misleadingly. There are many things you can do to guard against being taken in: understanding cognitive biases such as confirmation bias, the primacy effect and so on doesn't guarantee you won't exhibit those biases, but it can at least give you a warning about how to listen to any presentation of numbers and what pitfalls to look out for. Making sure you understand what baseline a statistic is being compared to, how the sample was gathered, what question the team that prepared the statistic were answering or asking: all of this can help to unmask persuasion, bombast and bullshit that is disguised as a logical argument.

In the end, whether you are on guard against simple unmotivated misuse of statistics or more sinister distortion, the basic lesson is that any statistic you are given could contain multiple flaws. It represents a sample of a population of some kind, as measured by flawed people who may be making mistakes in their methodology; the conditions that led to the results may already have changed over time; the data has been analysed in a particular way and the results displayed in a way that probably suits the motivations of whoever is displaying the results. So any single source of information on a topic should always be treated as nothing more than a starting point for discussion: you need to not only understand as much as possible about how that result has been reached and displayed, you also need, wherever humanly possible, to compare that result with alternative sources of information. Only by delving deeper and trying to take a broad view of any given situation will you truly be able to judge what is going on. So listen to statistics, but do so with great care and a sceptical ear and you may be able to discern the reality of the situation.

Bullet Point: As *The X-Files'* slogan famously claimed: 'The Truth Is Out There'. It just may not be the thing that people tell you first.

CHAPTER 3

The Corporate Numbers Game

THE MADNESS OF ORGANISATIONS

Some of the problems that beset politicians when it comes to targets and measurements of key indicators are also reflected in all kinds of organisations, from housing associations to councils, and from corporations to small companies. Bear in mind that the larger an organisation becomes, the more people within it are devoted to trying to protect the image of the company, which may or may not coincide with a drive to do the right thing in practice.

I have known a lot of people who worked in PR departments. ('PR' stands for 'Protecting Reputations' or 'Pulverising Reality' or something along those lines.)* There's usually nothing actually morally wrong with them. A large part of their role is just about trying to gain publicity for the company's products or services or to generate 'good news stories'. But a large part of their job does revolve around manipulating the truth, and the more dubious the practices of the company, the more duplicitous this can become, and the more they have to engage in doublethink to avoid feeling guilt or shame at what their job involves.

And it's not just PR workers who can become experts at doublethink. Many managers and directors have to simultaneously convince themselves that they care deeply about their employees' welfare while engaging in behaviour that is focused on making them work as hard as possible for minimal reward in order to maximise profitability.

The 2003 Joel Bakan documentary (and 2004 book) *The Corporation* argued that corporations are, by definition, deeply flawed. Since the US Constitution (and other legal frameworks) define the corporation as an individual, what sort of individual are they? The answer given is that it is

---

* OK, that was a lie.

one without conscience, which is legally constituted to escape personal responsibility and that feels no compunction in exploiting its employees, short-changing its customers and generally ignoring human rights and basic decency when the situation requires it. In short, the kind of individual they most closely resemble is a psychopath; and as we know, psychopaths are perfectly willing to engage in all kinds of deception in order to achieve their desired ends and goals.

In this chapter, I want to focus on some of the madnesses that organisations large and small impose on their employees, and, perhaps more crucially, some of the justifications and motivations that lie behind them.

The Enron Saga

As a parable of corporate deception there aren't many better examples than the story of Enron. The company was formed in 1985 from a merger between Houston Natural Gas Company and Omaha-based InterNorth Incorporated, with Kenneth Lay as CEO and chairman. The deregulation of financial markets had created a very different business landscape and Enron was repositioned as an energy trader and supplier, in which role it could now lay bets on future prices. Lay also created the Enron Finance Corporation headed by Jeffrey Skilling, one of the partners at McKinsey.

As the company headed through the nineties it appeared to be highly successful: indeed, it was named as America's most innovative company by *Fortune* for six years in a row. Bear in mind that the dotcom bubble meant that stock and asset prices were greatly overpriced in this period: Enron would not be the only company that would get into trouble because of excessive valuations. But where they stood out was for the sheer lengths they would go to to manufacture the appearance of success.

One of Skilling's earliest decisions was to change Enron's accounting practices from the historical cost accounting method to mark-to-market (MTM accounting). This means that assets and liabilities are valued by the company at their 'fair value' rather than at their 'actual cost' and are recorded at this value on the books. It can be a perfectly legitimate method, and came with Securities and Exchange Commission (SEC) approval (the US financial regulatory body), but Enron would abuse the method to an extreme degree. Once they started to log estimated profits as though they were actual profits, the company was starting to build an illusion of a company that was far bigger and more secure than the real thing. In particular, once they partnered with Blockbuster Video and entered the video-on-demand market, they started to log estimated profits that were wildly in excess of the final outcomes.

In 2000 the dotcom bubble started to rapidly deflate at a point when Enron was at its most overcommitted. One of their responses was to pour more money into high-speed broadband networks costing millions of dollars, which ended up being just more losses to add to the growing pile.

But as early as 1998, the company had been deliberately concealing its problems. When they bought an asset such as a power plant, they would immediately claim all the future profits on their books. If that profit wasn't realised, they took to concealing the losses by transferring the asset to special-purpose vehicles (SPVs): essentially off-the-books corporations that could be used to conceal the debts and keep them from Enron's books.

The scheme, which was masterminded by CFO Andrew Fastow, was even more complex: the asset would be exchanged with the SPV in return for cash or a note; Enron would transfer some of its own stock to the SPV, which could then use it to hedge one of the assets that were still on Enron's books. Meanwhile, Enron would guarantee the value of the SPV to avoid the problem of counterpart risk.

This resembles standard debt securitisation and was presented by Enron this way to its investors. But what they hid was the degree to which the SPVs were entirely orchestrated and created by Enron purely for the purpose of hiding losses.

Of course, knowing as we do that this story ended in disaster, we can see all too clearly why it went wrong. But we have to put ourselves for a moment into the shoes of the people at the top of the corporation who collaborated in and authorised all this deception. What on Earth were they thinking?

Well, firstly we have to allow for the conceit and self-regard of people who were being treated as stellar business successes, geniuses who were rewriting the rules. There must have been some degree of hubristic self-deception that prevented them from truly assessing their own actions. In addition, there was a sincere belief (shared with many at the time) that the steady growth in share prices was bound to continue. The people at the top of the company had become fixated on the share price, which reached a high of $90.75 in August 2000. But share prices are merely a reflection of underlying value and they were obsessing about the price itself rather than looking after the fundamentals that should have been justifying it.

And the belief that the share price would continue to rise must have contributed to the feeling that what they were doing wasn't really wrong. They probably saw measures such as the SPVs as temporary, clever solutions to the problem of shoring up the share price and avoiding the kinds of worries in the investing community that could have led to a loss of confidence. There are, of course, some ways in which many corporations do this: any kind of debt securitisation is, for instance, an attempt to make the company safer (and to make it appear safer). Managing appearances is a natural part of the job of the CEO and, to some degree, the CFO. Enron merely took this to an extreme, albeit one that was immoral, dishonest and dangerous.

In addition, there must have been other people at the company who were aware that the accounting practices were unorthodox and potentially unlawful. One of the factors that probably affected their decisions would be the power of peer pressure, the need to conform which corporations always stress, and the desire not to rock the boat: whistle-blowers may be key to uncovering instances of corporate fraud, but they rarely go on to be rewarded for their roles other than by losing their jobs and having to find a new career. Most employees of the company would have been reluctant to put their own heads on the block, so would have turned a blind eye or failed to really think their suspicions through. Lying by omission or by refusing to see the truth may be slightly less odious than out and out lying, but they still legitimise the deception and allow it to flourish.

It's also worth noting that the company's accounts were signed off as legitimate by one of the major accounting companies of the time: Arthur

Andersen. Their reputation took a major blow when they were found guilty of crimes with respect to their role (although a judge later overturned the convictions). Again, it is worth wondering what combination of self-deception, bombast and the desire to avoid rocking the boat contributed to the decisions made at a company that didn't have anything directly to gain (but had a lot to lose) from signing off such complex, devious accounts.

The Enron bubble truly burst in 2001. Lay retired early in the year and Skilling took over, before resigning himself for personal reasons. The share price halved over a few months as analysts started downgrading the stock. The company was forced to close one of its SPVs (known as 'Raptor') to avoid distributing nearly 60 million shares, but had to take the losses back on to its books: this finally triggered the SEC to get involved. Fastow was fired following an investigation and the company had to restate its earnings accurately, which revealed losses of $591 million and debt of $628 million. In December the company declared bankruptcy, with its share price down to $0.26. Its investors had lost $74 billion, of which just $21 billion would be repaid by the rump of the company following bankruptcy. Many of its employees lost their futures: both their jobs and their pensions had disappeared into thin air when the lies had come home to roost.

It is still one of the most spectacular company bankruptcies in corporate history and one that shows how deception can take hold deep within a corporation's practices once senior figures start to believe that the numbers are more important than the reality they are supposed to be measuring.

Bullet Point: If you run a company, look after the fundamentals, and let numbers like the share price look after themselves.

SHOULD COMPANIES BE MORE HONEST?

Let's first acknowledge that there are times when brutal honesty is not required in a commercial endeavour. Consider, for instance, the case of Gerald Ratner. A successful entrepreneur, he had built up the jewellery chain Ratners into a successful high-street business in the UK. It was regarded as a bit 'downmarket' but as being good at what it did until he put his foot in it with an unfortunate speech at the conference of the Institute of Directors at the Royal Albert Hall on 23 April 1991. He tactlessly said: 'We also do cut-glass sherry decanters complete with six glasses on a silver-plated tray that your butler can serve you drinks on, all for £4.95. People say, "How can you sell this for such a low price?" I say, "Because it's total crap."' He went on to describe one of the sets of earrings sold at the chain, saying it was 'cheaper than an M&S prawn sandwich but probably wouldn't last as long'.

After the speech was widely reported in the media, the chain immediately suffered a plunge in sales, and went into debt over subsequent years. It did eventually recover, under another name after several reorganisations, but Ratner's comments are remembered as a salutary example of why CEOs might not always want to be totally candid.

A similar incident led to the Swiss pharmaceutical maker Roche being listed in the '10 Worst Corporations of 2008' list compiled by Multinational Monitor. One of Roche's bestselling products at the time was the HIV drug Fuzeon. They had set a global floor price of $25,000 for a year's supply, bucking the usual trend whereby life-saving drugs are sold at lower prices in poorer countries.

South Korea's Ministry of Health, Welfare and Family Affairs attempted to put a ceiling on the price, valuing a year's supply of Fuzeon at $18,000. Roche would still have made a profit selling at this price, but chose instead to refuse the sale, meaning that many of South Korea's HIV patients had their treatment withdrawn. It's always worth bearing in mind that pharmaceutical companies are in the business of making money, not saving lives. This much is clear from their behaviour in many similar situations.

Bullet Point: We don't always expect total honesty from commercial organisations, but there are limits . . .

WHEN LIES BECOME THE TRUTH

It's also worth acknowledging upfront that there are potentially some positive reasons for people in business to lie. In particular, it is worth considering the **Pygmalion effect**. Studies in the classroom as long as fifty years ago established that if students are tested and then labelled as being likely to increase performance they are likely to perform better in subsequent tests for an extended period. Crucially, this happens even if the test results were completely fictitious.

Subsequent studies have shown that a similar **expectation effect** applies within organisations: when leaders communicate high expectations to individuals then they are more likely to work hard and persevere and to meet those expectations; and this is true even when the leader doesn't believe the message they are sending.

The expectation effect is similar to the placebo effect, in which doctors give harmless sugar pills to patients and assure them that they will help with their condition: the patient often shows clear medical improvement while on the medication in spite of the fact it has no medical impact. Both effects are self-fulfilling prophecies, in which communicating an idea or prediction produces behaviour that makes the idea or prediction come true. Companies like Amazon take this on board by letting their corporate employees know that they have very high expectations to live up to, and while some find this daunting or stressful, many push themselves to achieve more than they had imagined they could.

Think of a bank run: people believe the bank is going to fail, so rush to take their money out, leading to the bank failing. If the bank manager is able to persuade them that the bank's finances are sound, they don't rush and the bank doesn't fail. Businesses are often in a similarly precarious situation. They need investors, customers and employees to believe in what they are doing and to be convinced that the business will succeed. So even when the business owner or manager is in doubt privately, showing faith in their predictions of high sales, high turnover and rapid expansion can help to influence the situation: the sales person who pushes themselves that bit harder, the investor who leaves their money in the business, the customer who trusts that the warranty on a good will be honoured in future all help to make the numbers that were initially a lie become true.

The co-founder of Intel, Andy Grove, once said this to a business conference: 'Part of it is self-discipline and part of it is deception. And the deception becomes reality. Deception in the sense that you pump yourself up and put a better face on things than you start off feeling. But after a while, if you act confident, you become more confident. So the deception becomes less of a deception.'

Bullet Point: Sometimes there may be good reasons to lie.

WHY PEOPLE LIE IN EVERYDAY SITUATIONS

Before continuing to talk about corporate lying I want to digress for a moment to consider some of the reasons why people lie. From little white lies to huge deceptions, lying is pretty common: there genuinely has been some research into this. Twenty-something years ago, for instance, Bella DePaulo, a social psychologist at the University of California, Santa Barbara, asked subjects to keep a diary of the times they had intentionally tried to deceive someone: in this particular experiment they found that subjects lied one to two times a day on average, ranging from innocuous lies like trying to protect someone's feeling or cover up a minor oversight through to major whoppers such as lying on a job application or about an affair.

In spite of this, human communication proceeds on the basic, mostly reasonable presumption that people are more likely to tell the truth than to lie: our predisposition is to trust that people aren't lying on most occasions, even if we aren't naïve. Professor Timothy Levine, a professor at the University of Alabama who has been researching into human deception for twenty years, calls this **truth default theory**. And if people are lying on average a few times a day, then clearly truth-telling is indeed more common than lying.

However, people do lie. One analysis Levine gave of the most common reasons for lying produced the following list:

Personal Transgression (covering up a mistake or misdemeanour) 22 per cent

Economic Advantage (to make money or gain financial advantage) 16 per cent

Personal Advantage (to gain other types of advantage) 15 per cent

Avoidance (escaping or avoiding other people) 14 per cent

Self-Impression (to shape or protect our self-image) 8 per cent

Unknown (we don't know the reasons ourselves) 7 per cent

Altruistic, Social or Polite lies 7 per cent

Humour (to make others laugh) 5 per cent

Malicious lies (designed to hurt others) 2 per cent

Pathological lies (in denial of reality) 2 per cent

Obviously, within the workplace several of these, in particular the first two, would be common reasons why people lie. But I'm more interested here in why organisations end up being deceptive. So the more crucial question is how those lies operate, and here I think we need to look at two further types of lying (which would in some cases cross over with the list above).

Bullet Point: We expect people to tell the truth, but lies are still common.

LYING PSYCHOPATHS

Having mentioned the concept that corporations can behave in ways that resemble a psychopathic individual, we need to pause briefly on the question of how often that is because the individual employees of that corporation are **actual psychopaths**.

Well, it has commonly been observed that psychopathic traits can be an advantage when it comes to roles such as being a politician or CEO: psychopaths can certainly succeed in business, partly because they lack empathy, a conscience and a compunction about lying for personal advantage, since selfishness is their fundamental motivation.

There is some dispute about what motivates lying in a psychopath. On the one hand, many psychopaths believe that whatever claim they are making at a given moment really is true. On the other hand, psychopaths can use lying as a means of control: there are some psychopaths who seem to get a thrill out of lying as obviously and openly as possible, then becoming outraged and aggressive when those lies are questioned. Possibly the moment when they are questioned is precisely when they come to believe their own version of reality, even if the lie started out as something malicious or controlling.

Above we noted that a business leader will sometimes project confidence they don't truly believe in so as to create a self-fulfilling prophecy. The truly psychopathic CEO probably doesn't fall into this category: for them the lying is either complete denial or reality or a game that they are playing.

A genuine psychopath is a very difficult person to communicate with, but the good news is that while they make up a relatively high proportion of criminals, they only make up about 1 per cent of the general population. So while psychopaths may in some cases be responsible for corporate lies, which they will bully others into complying with, we also need to look at what is going on with the 99 per cent of the population who aren't such natural-born liars.

Bullet Point: Genuine psychopaths lie because they only care about themselves and believe reality will bend to their will. Your boss might be one, but you probably aren't.

The Psychology of Self-Deception

To my mind the most interesting area to investigate when it comes to the psychology of lying in organisations is the concept of **self-deception.**

There is a divergence between the fields of psychology and philosophy when it comes to explaining self-deception. Firstly, while this is a simplification, traditional philosophy tends to make two assumptions: people have a single, coherent self and are rational beings.

On these assumptions, how can self-deception be analysed. Firstly, let's see what is going on when person A deceives person B into believing that x is true.

A knows that x is false. A tells B that x is true. B thus comes to believe that x is true.

Now, can we adapt this train of thought for self-deception?

A knows that x is false. A tells A that x is true. A thus comes to believe that x is true.

This seems like a straight-up contradiction since it implies that A believes both of the statements 'x is true' and 'x is false'.

At this point there are two ways that philosophy can go. One is to query the idea that the self is indeed a single, unitary thing. Philosophers such as David Hume have toyed with the idea that the self is a myth and that we are merely a 'bundle of ideas', but while this is an interesting concept it doesn't help us that much when it comes to explaining self-deception. The second train of thought that some philosophers have pursued is to explore the idea that we might deceive ourselves but that it might not be intentional.

But rather than getting too bogged down in that debate, it might make more sense to turn to psychology. Since the time of Sigmund Freud, the idea that we have conscious, subconscious and unconscious mental process has been widely understood as a good basic model of human behaviour. This leads on to the idea that we often behave in particular ways because of unconscious processes, but rationally explain these through conscious thought. Thus we sometimes believe we are thinking rationally when we are merely rationalising after the event. This would be one way of explaining why we may have quite self-contradictory aspects to our behaviour and communication.

There is also a biological angle we can take on this subject. The evolutionary biologist Robert Trivers has explored the role of deception in animal behaviour and evolution. From alarm calls to mimicry, deception is a part of the animal world that predates even the evolution of language. So one possibility is that we learn to deceive ourselves in order to better deceive others. When we lie consciously, we show signs of this in our physical traits: our eyes might dart around or our nostrils flare. The more we actually believe what we are saying, the less visible these giveaways will be. And if this ability bestows an evolutionary advantage, then natural selection would mean that the human species become more adept at self-deception over time.

There might or might not be something in this. But I think the psychology of self-deception can perhaps be explained relatively simply, once we accept the psychological concept that we don't always know ourselves or understand our motivations and combine this with the concept of cognitive dissonance, whereby we find it jarring when new information clashes with something we already believe and feel the need to eliminate or undermine one belief or the other. (This accepts the philosophical idea that we don't generally consciously hold two opposing beliefs, since we find it uncomfortable to do so, but still allows us to see that our unconscious motivations may lead us to dismiss or rationalise the facts we are uncomfortable with.)

For instance, we tend to think of ourselves as being fundamentally good people. But we sometimes end up doing bad things, such as lying for our own sake or for the sake of our job or company. The idea that we are a good person is more deeply ingrained and more precious to us than the idea that we are doing something bad. So we have a strong motivation to rationalise what we are actually doing, blame others for it, belittle its significance, or simply refuse to think too hard about it.

So, when a wife or husband is having an affair, they may come to blame their spouse or to believe it is not as big a deal as the spouse would think. They will focus on good times in their marriage and think, 'See: that's who I really am.' A salesperson who is selling a toxic product may persuade themselves that it isn't really doing any harm, or that their customers don't deserve any better, or in some other way protect their self-image.

My argument would be that self-deception is often present when organisations are involved in bad or deceptive behaviour. Psychopaths may be involved at some stage of the process. But most of the people carrying out the work come to believe they are simply 'doing their jobs' and that if there are bad consequences, it's just not their fault.

Bullet Point: Self-deception is, essentially, what we engage in when 'what we actually do' conflicts with 'who we like to think we are'.

MISSION STATEMENTS: WHAT ARE THEY GOOD FOR?

Now, let's take a moment to think about mission statements. In the 1950s and 1960s, most companies managed without a mission statement. After all, their fundamental purpose was to make money, by giving their customers a product or service they were willing to pay for. They didn't feel the burning need to tell people that what they were actually doing was making the world a better place.

Since Peter Drucker's day (see p. 25) the practice of scientific management has evolved into the entire industry of management consultancy. Management is increasingly fetishised as a heroic endeavour and this has arguably fed into the more bombastic approach to mission statements that has taken hold since about the 1980s.

Now, most companies, even down to SMEs (small to medium enterprises), tend to feel obliged to have a mission statement of some sort. And note that many mission statements start out by stating 'Who We Are'. So the mission statement is essentially the company's self-image: it is how they would like the world to see them.

For instance, Enron hilariously had the motto 'Respect, Integrity, Communication and Excellence'. Its 'Vision and Values' mission statement included the words, 'We treat others as we would like to be treated ourselves … We do not tolerate abusive or disrespectful treatment. Ruthlessness, callousness and arrogance don't belong here.'

So, this is who they liked to think they were. But there was a clear disconnect between this belief and what actually happened.

This is, essentially, self-deception on a corporate level. It might have been sustained by some psychopathic behaviour, but there would also have been individuals who protected their own self-images by choosing to believe that Enron's financial practices were normal and acceptable.

So, from the top down, self-deception helps to fuel organisations, especially when they are failing to live up to the image that is presented in their mission statements. There are, of course, plenty of genuinely ethical, decent companies around. But since this is a book about lying, the ones we are more concerned with are the ones whose mission statements don't conform to their real-world behaviour.

Bullet Point: If a company or organisation has let you down, it's always worth checking their mission statement if you want a hollow laugh.

CUSTOMER CARE TARGETS

The mission statement of a company or organisation will often include commitments to customer care. These are fascinating because they usually seem reasonable but allow enough wiggle room to entirely undermine the concept of customer care. (And this is even before you take into account how often their measurement of this target is based on five-point surveys – see p. 30 – and similar shallow metrics.) I'll give a few anonymous examples of company behaviour around the world to emphasise the ways in which companies actually behave.

One of our friends has a flat for which the maintenance is overseen by a housing management company. They have a grand head office, a mission statement and a long list of customer care promises. This includes a commitment to respond to all repair requests within forty-eight hours, to respond to emergency call outs within four hours, and to publish a log online of all repairs and their resolutions.

This all sounds good. Until you try to call to report a repair and are kept on hold for two hours. Then you realise that simply returning a phone call to tell you an appointment has been made counts on their log as 'responding to a repair request'. The actual appointment for a worker to call may take weeks. Often the worker doesn't show up but the housing company still claims they did. Then, when it finally happens, the first visit is often from someone wearing a suit and carrying a torch and clipboard rather than dressed in work clothes: they will simply look at the problem (whether it be a leaking toilet, missing chimney pot or broken light fitting) and make a note of it on said clipboard. At this point the repair is often flagged up online as 'resolved' and it takes more phone calls to provoke further action. The further calls are logged as new repairs rather than as the same old repair, thus keeping average repair times down, which makes them look better in the company's annual report.*

Similarly, emergency calls (such as water pouring through the ceiling and flooding the bedroom) really are responded to in four hours. But the first visit is from someone who does the least work possible (turning off the water supply) and disappears, leaving the apartment waterless for an unknown amount of time.

The infuriating thing in all these cases is that the repairs are logged as resolved or omitted from the online log, meaning the annual report shows constant improvement in the repairs service at the same time as there are fewer and fewer operators on the phone lines, and more and more attempts are required to achieve a successful outcome. The company slaps itself on the back, and its self-image and numbers are preserved, but for the tenants the customer care is just dehumanising and bewildering.

Bullet Point: Customer care targets are often deeply misleading.

Known Problems

Here's another way that companies can pretend they are respond-ing to complaints without really doing anything. My wife once reported a problem with the service to her mobile phone company. They responded within twenty-four hours with a message saying that it was a 'known problem', they were working on it and would let her know when it was resolved. She never heard from them again and the problem remained unresolved when she closed her account in favour of another company two and a half years later.

---

* One of our local London councils appears to be behaving in a similar way with their pothole repair service. When a pothole is reported, their response is to send out a worker, who carefully paints around the pothole in white paint, after which it can take months before anything is actually done to fix the problem.

PRETENDING TO SAVE THE PLANET

Of course, there are many different ways of using numbers to pretend to care for your customers. Let's revisit the Volkswagen diesel testing scandal, which is another example of how deception can take deep root in a corporation, but which is also a story that ends with a bit of hope for the future.

The overview is that Volkswagen's reputation was damaged when it was revealed that it had been using a piece of software that could fool the emissions testing regime: it could tell when the car was being tested and would temporarily reduce its harmful emissions, allowing the car to pass the test even though it would not do so in ordinary driving.

The story starts in 2008, when tougher emissions standards were brought in for new cars. Most of Volkswagen's rivals used a system called AdBlue which is a urea-injection system that uses a chemical catalyst to avoid unburnt fuel getting into the exhaust component. However, Volkswagen claimed they would be able to meet the new standards in their diesel cars without using AdBlue. While there is no confirmation of this, it may have been a cost-driven decision since the AdBlue system was pricey.

For years, their tests suggested that they were indeed managing to achieve this promise, and this contributed to the widespread belief that diesel was becoming a greener, cleaner fuel than petrol. But in 2013, the International Council on Clean Transportation collaborated with West Virginia University for a study on the Volkswagen diesel cars' emissions levels. They weren't trying to catch VW out; in fact, they were only studying VW cars as they saw them as excellent examples of how to build a clean modern car. They ran the emissions tests both in real-world driving conditions and in a laboratory. They were startled when the results in the two environments showed a massive divergence in the VW cars. The Jetta they tested exceeded the emissions caps by fifteen to thirty-five times, while a Passat exceeded them by five to twenty times.

In 2014, they alerted the California Air Resources Board and the Environmental Protection Agency, who asked Volkswagen for an explanation. Initially VW disputed the findings, citing 'technical issues', but they did recall half a million cars for a software patch. However, that patch didn't fix the problem for the cars it was installed in and eventually the true cause was discovered: a piece of software called the 'switch'. The inputs to the switch were the position of the steering wheel, speed of the wheels, length of the 'journey' and barometric pressure. It could temporarily cut emissions when these inputs matched standard testing conditions. So the switch was there in order to output the correct numbers but, crucially, only in test conditions.

It was software designed to output a lie.

When VW was presented with evidence of the switch they finally admitted that it had been a deliberate scheme to lower the results of their **emissions testing.** Their shares immediately fell 20 per cent and then fell further when VW admitted that as many as 11 million cars could be affected, meaning that tens of thousands of excess emissions had been released into the atmosphere in the US alone, with much more having been released around the world. CEO Martin Winterkorn issued a belated apology: 'I personally am deeply sorry that we have broken the trust of our customers and the public.' VW set aside €6.5 billion to 'cover the necessary service measures and other efforts to win back the trust of our customers'. And eventually Winterkorn was obliged to resign.

As with other corporate deceptions discussed in this book, one wonders how many people were involved in this particular deliberate deception and how they justified it to themselves and to their children. But it is the younger generation that provide a moment of possible light at the end of this particular tunnel.

Writing in the *Harvard Business Review*, Michael Schrage has argued that the combination of citizen science and ethical consumerism may make it harder and harder for corporations to behave as VW did in future. It is becoming more and more feasible to distribute home testing devices using social media, Kickstarter campaigns and other distributed methods, and these can be used to keep manufacturers to their promises. After all, VW were under no particular suspicion, but it was an organisation that cared about clean air that ended up uncovering their malpractice.

Schrage suggests that: 'As social media, disposable sensors, smarter phones, machine learning platforms, savvy consumer activists, self-quantification and the "internet of things" accelerate into the economic mainstream, betting billions on the stupidity of one's customers becomes a fool's errand.'

Bullet Point: Dare we hope that future corporations will learn that it is much, much safer to keep their ecological promises rather than just pretending to do so?

THE PINTO CASE

The story of the Ford Pinto is a fascinating case study in business ethics. Some have described it as demonstrating the way that corporations place a cheap value on human life and gamble with it, while others have seen it as a case of public misperceptions of the decisions that all manufacturers have to make.

The Pinto was a small car, sold from 1971 to 1980 in North America, produced in response to the growing popularity of smaller cars from Europe and Japan in the American market. Planning for the model started in 1967: Ford President Lee Iacocca had mandated a model that would weigh less than 2,000 pounds and would be priced under $2,000 (so that it could be marketed as costing a dollar a pound). The planning schedule was unusually fast, and decisions that might threaten the schedule were reportedly discouraged.

During the design process it was decided to place the fuel tank behind the rear axle rather than above it, in order to leave a bit more room for the boot, which otherwise would have been very small. This meant that the fuel tank was vulnerable to a rear-end collision. There were several factors that added to the problem: there was only nine inches of 'crush space' between the axle and the fuel tank. And the Pinto's rear bumper was ornamental and not well fortified: many other models of cars had horizontal and vertical components known at 'hat sections', which stopped bumpers from crumpling too easily. Finally, there were protruding bolts in the rear bumper that could puncture the tank in a collision and the fuel filler pipe was designed in a way that made it vulnerable to becoming detached in a collision. Oh, and the doors had a tendency to jam after collisions, which risked passengers being trapped in a burning vehicle.

Now, before continuing, let's acknowledge that all cars, especially cheap models, come with some design compromises, some of which are more problematic than others. It has been pointed out that, while the Pinto had a worse record for fires involving rear-end impacts, these are only a subset of accidents and the safety record of the model when it comes to 'all fires' was more or less average for compact cars.

However, the problem with rear-end collision fires soon started to bring the company unwelcome publicity. In 1972 a Pinto driven by Lily Gray stalled on a California freeway, before being hit from behind by a car that had decelerated from 50 mph to about 30 mph. Gray died in the resultant fire while her passenger, Richard Grimshaw, suffered serious burns. This led to the case of Grimshaw v. Ford Motor Company.

A California court awarded $2.5 million in compensation damages, plus $3.5 million in punitive damages. To make things worse for Ford, Grimshaw's lawyers shared information about the case with the investigative magazine *Mother Jones* and The Center for Auto Safety. The latter persuaded the National Highway Traffic Safety Administration (NHTSA) to launch an investigation into the Pinto, although they decided not to demand a recall in 1974.

The problems continued, however. Estimates of deaths linked to the vehicle have varied: one official estimate was that there were twenty-seven such deaths. However, in 1977 *Mother Jones* published an article that claimed that the number of deaths was in the hundreds. It claimed that Ford had lobbied against safety standards that would have forced them to make changes to the Pinto and that they had stalled for eight years over meeting new suggested standards. Most damningly, it argued that Ford had done this because 'its internal "cost-benefit analysis", *which places a dollar value on human life,* said it wasn't profitable to make the changes sooner'.

That dollar value was $200,000. (The $200,000 figure actually came from the NHTSA but it was prepared in collaboration with the industry.)

Now, this is the part of the Pinto story that still resonates, several decades later: the idea that companies can secretly put a value on your life and then decide whether it is worth their while to spend a bit more money to save it. So it's worth exploring what actually happened.

In 1973 Ford's Environmental and Safety Engineering division prepared a cost-benefit analysis called *Fatalities Associated with Crash Induced Fuel Leakage and Fires*. It submitted this to the NHTSA in support of the company's objection to tighter controls on fuel systems. This document came to be known as the 'Pinto Memo'; in it the engineers gave an estimate for the cost of the modifications that would be required to reduce fire risk at $11 per car which, given that there were 12.5 million cars in total on the road that would need those modifications (including other manufacturers' cars), would make a total of $137 million. They also estimated that this would save 180 deaths and 180 serious injuries in fires per year, meaning that the 'cost to society' would be lower than the costs to the industry of the modifications.

This was treated as the smoking gun in the Grimshaw case as it showed the company coldly comparing their own potential costs to the 'costs' to the victims of rear-end fires.

Ford did eventually issue a voluntary recall in 1978, taking millions of vehicles off the roads, but it was widely seen as a belated gesture. A large part of the problem was the way that Ford dealt with the PR issues involved with the Pinto: rather than directly confronting the issue, they were busy issuing denials and fighting against claimants. After the event, Lee Iacocca was quoted as saying that, 'Clamming up is what we did at Ford in the late '70s when we were bombarded with suits over the Pinto ... The suits might have bankrupted the company, so we kept our mouths shut for fear of saying anything that just one jury might have construed as an admission of guilt. Winning in court was our top priority; nothing else mattered.'

Bullet Point: In inflation-adjusted figures, your life is worth about $830,000.

Cost-Benefit Analysis

The Pinto case is pretty depressing with regard to what it says about business ethics. But I'm going to say a couple of things in defence of Ford. It is inevitable that most people have a gut reaction to the way that their engineers placed a value on human life, but the former UCLA law professor Gary T. Schwartz (among others) has argued that the Pinto case has acquired an emotional, or even mythical, charge that leads people to misperceive the basic ethics that are at work.

It is clearly impossible to make a product that is 100 per cent accident-proof, especially one that is likely to be moving at high speeds. You could theoretically make a car safer and safer by spending more and more money on safety features but you are still dealing with unknowns that include the behaviour of drivers, and acts of God such as falling trees or lightning strikes. But the more expensive it is to produce a car, the fewer people will be able to afford to buy it. So where do you draw the line when it comes to incorporating safety features, especially in cheaper models?

There may be no perfect answer to this question. But cost-benefit analysis is one of the tools that industries use when they try to deal with difficult ethical decisions. And there is a sense in which this is inevitably going to result in a kind of deception: no car manufacturer would want to advertise a car by telling you which safety features they had left out, how much they would have cost, and what calculations they had made in order to make this decision. And the car industry is far from being the only industry that has historically lobbied government bodies to try to limit safety regulations that would increase their costs.

So while Ford's behaviour over the Pinto case was pretty poor, it is an example of a basic problem in business ethics. However, it is worth focusing for a moment on their use of cost-benefit analysis, which has an intriguing history.

As a formal tool, it only dates back to 1848 when the French engineer Jules Dupuit wrote an article in which he calculated 'the social profitability of a project like the construction of a road or bridge'. This was rooted in the economic concept of 'utility' (see p. 172), which he suggested could be measured by finding out how willing people would be to pay for something. By adding up the total estimated utility of everyone who would use that road or bridge, the government could estimate whether it was worth proceeding or not.

The first real application of cost-benefit analysis (CBA) in the US came in the Flood Control Act of 1939, which brought control of local and federal dam projects under the control of the United States Army Corps of Engineers. So from the earliest days of the use of CBA in the US it was being applied to decisions that could potentially be the difference between life and death. In the 1960s, President Kennedy's secretary of defense, former Ford Motor Company president Robert McNamara, was a CBA zealot, and in this period it was increasingly applied to decisions over public policy on mental health, chemical waste, education, drug abuse and healthcare.

Similarly, in other countries, cost-benefit analysis has come to be a large part of policymaking, which often applies to decisions that could affect lives. Health insurance companies around the world place a value on a year of human life, whether this is openly explained or not (a recent estimate of this suggested that most insurance companies use a ballpark figure of $50,000 a year). In the UK, the National Health Service makes the same calculation: the NHS drugs watchdog, the National Institute for Health and Care Excellence (NICE), uses a method called 'QALY' – the quality-adjusted life year measurement – to assess whether or not a particular treatment is worth using their resources for.

A QALY is an estimate of how many extra months or years of 'reasonable quality' life a patient can expect to get as a result of a treatment. And this has to be assessed against the costs: a NICE spokesperson has explained that 'Each drug is considered on a case-by-case basis. Generally, however, if a treatment costs more than £20,000–£30,000 per QALY, then it would not be considered cost-effective.' This means that drugs that might extend someone's life can be unavailable due to cost.

Cost-benefit analysis is essentially a product of our age. Since the nineteenth century we have mostly lived in a capitalist system, in which profits tend to dictate what is feasible: some might think that numbers should never be more important than human life, but that isn't the world we live in.

In addition, as we have been able to achieve more and more technological and medical marvels, there are genuine problems about how to deal with the ethical dilemmas that this raises.

How you feel about all of this may depend on your political standpoint. But the key point is that there is an inevitable degree of doublethink involved when an organisation uses cost-benefit analysis regarding life or death issues: no one wants to tell you to your face how much they think your life is worth; instead they want you to believe they are on your side. Exactly how they resolve that conflict is a matter of tact, policy and expediency.

Bullet Point: It's not just car manufacturers who calculate how much your life is worth.

Coronavirus Update

Here's a story about cost-benefit analysis. In 2016 an advisory group set up by the UK Department of Health and Social Care recommended stockpiling personal protective equipment (PPE) for use in a potential future pandemic. The recommendations included 'providing eye protection for all hospital, community, ambulance and social care staff who have close contact with pandemic influenza patients' because of the risk of infection via contact. However, in June 2017, a department official objected to 'the very large incremental cost of adding in eye protection', noting that it would 'substantially increase the cost of the PPE component of the pandemic stockpile four- to six-fold, with a very low likelihood of cost-benefit based on standard thresholds'.

The result of this decision and various other ways in which the pandemic stockpiling was watered down is evident as I write. Medical professionals who are putting their lives on the line are being left without adequate PPE, while ministers perpetually repeat bland statistical assurances that we will soon have enough, that shipments are on the way, that so many million pieces of PPE have been distributed and so on.

Right now, there are heroic doctors and nurses fighting for their lives because someone decided it would cost too much to protect them.

Big Pharma

When insurance companies or bodies such as NICE calculate whether or not it is worth saving your life, there are of course two sides to the equation. On their side there is the decision as to how much to spend; on the other side it is the drugs companies who are setting the prices of the drugs.

This is a complex question: drugs companies do have to spend money to develop and test new products, and they have limited patent periods to maximise their revenue from a new treatment, after which other manufacturers can, in theory, make generic versions so, unless they have managed to position their version as being superior or to develop a version sufficiently different to warrant a new patent, they lose their competitive advantage. That is just the system in which they are operating, so let's not pretend they don't face some significant challenges in the market.

But . . .

The pharmaceutical industry sells to a market that is worth more than $300 billion, mainly funded by private or public healthcare. It is a market that is dominated by a small number of corporations who spend a lot of time lobbying regulators to legislate in their interests. The cost of new drugs keeps rising and the number of new products is gradually decreasing.

And there are significant questions about the way their drugs are priced. There are numerous cases in which the prices of the drugs have been dropped to affordable levels only after public pressure. For instance, the breast cancer drug Palbociclib was originally rejected for use in the NHS because it was estimated to cost about £140 a pill. Following a petition from sufferers and sympathisers it was finally authorised for use after Pfizer dropped its price to an undisclosed level. And this is only one example of flexible pricing. There have been numerous cases in which the UK Competition and Markets Authority has fined various drugs companies for unfair pricing of medicines (and similar cases have been seen in other countries). Pfizer, along with a smaller UK company called Flynn Pharma, was fined £90 million in December 2016 for 'excessive and unfair' pricing of an anti-epilepsy drug. Meanwhile, GlaxoSmithKline and two small firms were fined £45 million for conspiring to delay competition on an antidepressant.

Also, the generic competition to a drug doesn't always materialise when the patent period ends. In some cases deregulation actually leads to huge price hikes: at one point, Concordia's Liothyronine tablets were priced at less than £5 but following debranding they increased to over £250 by 2017.

There have been claims from the industry that it costs them up to $2.6 billion to develop and market a new drug. But this is a debatable figure. One analysis of new cancer drugs published in *JAMA Internal Medicine* estimated the median cost of developing them much lower at $648 million. A fairly large proportion of development money comes from governments in the first place. And a 2015 US investigation found that nine out of ten of the biggest pharma firms were spending more on marketing than they did on research.

Thus far, it is all a bit murky. The pharmaceutical companies are vigorous in defending their activities, and we can all agree that some of their products have been of great benefit. But there are some much more questionable aspects to the industry.

America in particular is currently in the grip of a severe opioid addiction epidemic. There has been a huge level of overprescription of certain drugs that are extremely addictive. Almost 400,000 people have died from overdoses involving prescription or illicit opioids over the past two decades, according to a government body (the CDC). In 2019 alone there were several major cases in which companies and individuals were convicted of wrongdoing with respect to these drugs.

For instance, John Kapoor, the founder of Insys Therapeutics, was found guilty of criminal conspiracy, together with four of his fellow directors. They had bribed health professionals to prescribe their fentanyl spray Subsys, a painkiller that is up to 100 times stronger than morphine. The bribes added up to more than $10 million. The doctors had been encouraged to recommend Subsys to people who didn't need it and also to increase their doses. So essentially they were not just drug pushers, but corrupt drug pushers.

Meanwhile, the world's largest healthcare company, Johnson & Johnson, were ordered to pay $572 million by an Oklahoma judge who ruled that they had caused a 'public nuisance' and had been a significant cause of the opioid crisis in the state, which has suffered severely from it: for instance, in 2009, drug overdose deaths surpassed car-crash deaths there, and in 2012, there were 128 painkiller prescriptions per 100 people.

The case was based on the claim that J&J had been mis-marketing its own opioid products while supplying the raw materials to other opioid makers (two of whom also settled in the same case, although without admitting liability). Their sales representatives had been trained to distribute the message that pain was being undertreated and to downplay the risks of prescribing opioids. In order to convince medical professionals that they weren't doing enough to help people in pain, they were coached to use the description 'pseudoaddiction'* to describe patients that were 'presenting the symptoms of addiction'.

Addicts, in other words.

Bullet Point: Remember what the man said: it's about making money, not saving lives.

THINGS THAT DEFINITELY AREN'T ADDICTIVE

One of the major challenges of the pharmaceutical industry is to come up with medication that continues to be effective without becoming addictive. This is a difficult challenge, for natural medical reasons. But a truly non-addictive sleeping pill or painkiller that had no harmful side effects would be the goose that lays the golden eggs.

The problem is that most products that are originally hailed as being the new, safe option turn out to have problems over time. The explanation is a simple question of numbers. When you test a new drug, you do it in limited conditions for a limited amount of time. A drug for which no withdrawal symptoms are observed after a few weeks may

---

* Incidentally, when patients who have been on antidepressants suffer withdrawal symptoms, the pharmaceutical industry prefers to describe it using the more anodyne term 'antidepressant discontinuation syndrome'.

nonetheless turn out to be severely addictive after it has been prescribed for a year or more.

The first benzodiazepines (or 'bennies') came on to the market in the early 1960s and were initially treated with great enthusiasm in the medical sphere as they were less prone to causing overdoses than their predecessors, the barbiturates. However, it soon became evident that drugs such as Valium (the brand name for diazepam) were addictive, and difficult to quit.

A similar pattern has been repeated with opioids. For instance, Purdue introduced the drug OxyContin (a brand name for oxycodone) in 1996, making the bold claim that it could relieve pain for twelve hours. Purdue's marketing of the drug quoted two studies that showed that the risk of addiction was less than 1 per cent, but these were not from long-term users. Studies on longer-term use of opioids suggest that 3–17 per cent is a more plausible range. When doctors reported back that patients were not getting the promised twelve hours' dose, Purdue (which had known of this problem since the launch) advised upping the dose rather than prescribing more regular, smaller doses. This was allegedly at least partly because they wanted to maintain their claim of twelve-hour relief for marketing reasons.*

Oxycodone has now become America's most used recreational drug. And this is dangerous as Oxycodone users suffering withdrawal symptoms often turn to heroin, as it is cheaper. In May 2007, Purdue pleaded guilty to misleading the public about OxyContin's addiction risks and were ordered to pay $600 million. There are still many outstanding cases against them and they were one of the companies that settled in the Oklahoma case I mentioned. And in September 2019 they filed for Chapter 11 bankruptcy.

Bullet Point: In the future there may really be sleeping pills and strong painkillers that aren't addictive. But they aren't here yet, and any claims to the contrary are based on flawed numbers and logic.

* https://www.latimes.com/projects/oxycontin-part1/

The Problem with Thalidomide

While we're on the subject of drugs that turned out to have side effects, the story of thalidomide is an especially sad one. The drug is still used today to treat various cancers and HIV symptoms, but in a highly controlled way. But it was originally marketed much more loosely for conditions including anxiety, sleep problems, stress and, crucially, morning sickness. The animal testing carried out on the drug hadn't included pregnant animals and, in spite of existing evidence that alcohol drunk during pregnancy could affect the foetus, it was wrongly assumed by many pharmaceutical professionals that drugs in general were not absorbed through the placental wall. As an example of the marketing of the drug, the UK brand (known as Distaval) was marketed with this information: 'Distaval can be given with complete safety to pregnant women and nursing mothers without adverse effect on mother or child . . . Outstandingly safe Distaval has been prescribed for nearly three years in this country.' Abnormalities including deformities and missing limbs were found in many children of mothers who had purchased thalidomide over the counter. The drug was mostly withdrawn or heavily regulated from 1961 onwards, but by then up to 20,000 babies worldwide had been affected, some with severe disabilities. The scandal that followed did lead to a significant tightening up of the testing regime, and to a renewed focus on regulation of the pharmaceutical industry, but that wasn't much consolation to the many families who were affected.

MINIMAL MANAGEMENT

I mentioned earlier Peter Drucker's comment that 'What gets measured gets improved' and the target culture that has arisen in management and politics. It is worth mentioning that many industries now rely on metrics and algorithms rather than man management. One of the reasons why this is appealing is that it allows for far fewer people in management. This has especially been the case in internet companies, who work on tight margins and often have to upsize from small start-ups to global level while maintaining the business model developed at the small scale.

As an example, Amazon is one business that has expanded to the point that it is worth hundreds of billions of pounds. It is highly reliant on data and algorithms through its entire operation: at a corporate level it uses a software system known as the Anytime Feedback tool,* whereby employees can share praise or criticism of one another's performance.

At the company's huge warehouses from which the goods are dispatched, there is an algorithm that locates an item and sends a picker to fetch it. The picker scans it using a handheld device and as it progresses through the fulfilment centre it is continually monitored. And there are similar systems in place to keep track of the performance and position of delivery drivers, whether they are working directly for Amazon or for other proxies.

This system also allows for constant monitoring and evaluation of each employee's performance and breaks. The workers have productivity targets, which again are monitored by software. And, as with other high-tech companies, there is ongoing work on replacing at least some parts of this chain with automated systems, whether it be robotic vehicles to shift stock around the warehouse, delivery by drone or whatever. Robots are expensive in the short term, but potentially much cheaper in the medium to long term.

Amazon is one of many of the new breed of companies to rely on zero hours contracts: many of the employees are effectively treated as self-employed. In many companies algorithms are used to allocate the work to employees on zero hours contracts, meaning that people don't know from day to day how much they will be earning that week. In many cases there is little transparency about how the algorithms work (and even the management may not fully understand them: as AI and machine learning is increasingly a part of the high-tech scene, so management algorithms may be making decisions that no human has played a part in).

* https://www.bbc.com/future/article/20150818-how-algorithms-run-amazons-warehouses

The gig economy in general is also reliant on zero hours contracts and algorithms. Consider the case of Uber. The drivers' jobs and routes are allocated by an algorithm that then tracks their compliance with the request, using GPS to monitor speed and position along the way. Drivers' performance is also monitored by the customer feedback mechanism, which also feeds into the algorithms that allocate work. The drivers often have little understanding of exactly how the jobs are allocated, which can lead to a feeling of alienation. Uber has also admitted to using techniques derived from behavioural science to 'nudge' the drivers to, for instance, work longer hours.*

It's no secret that some employees of large corporations who work in similar conditions to those at Amazon and Uber have expressed their unhappiness with the way they are managed. So this provides a challenge to the corporations who naturally want to project themselves as good employers and to motivate people to work for them.

Let's just look at one documented case in which misleading figures were used to this end.† In 2017 Uber were chastised by the Federal Trade Commission (FTC) in the USA for advertising jobs using unfeasible estimates of the income they would be able to earn. Uber offered the settlement without admittance of wrongdoing, but the money was intended to compensate drivers who had been convinced by ads promising income that was tens of thousands of dollars higher than the money they were actually able to earn, and which didn't appear to figure the realistic operating costs they faced.

According to the FTC, Uber's website claimed that UberX drivers in New York and San Francisco earned median incomes of around $90,000 and $74,000, which were overestimates of $29,000 and $21,000 respectively. The FTC also criticised the company for overstating drivers' hourly rates in ads on websites such as Craigslist, and for understating the likely costs of leasing a vehicle from the company's Vehicle Solutions Program.

* https://www.nytimes.com/interactive/2017/04/02/technology/uber-drivers-psychological-tricks.html
† https://www.forbes.com/sites/janetwburns/2017/01/23/uber-must-pay-20m-for-luring-drivers-with-inflated-wage-stats/#eb7594a191d2

As a result Uber was instructed to keep its advertising claims about wages and costs factual and some drivers received welcome compensation packages. Of course, in the long run this will be less of a problem for a company such as Uber. It is heavily invested in self-driving cars (known as autonomous vehicles or AVs), so eventually drivers will be unnecessary and the algorithms and the robots can continue to interact with each other without humans being required at all.

Bullet Point: Algorithms and robots really are taking over the world.

CONCLUSION: WHEN COMPANIES LIE

There are a lot of obvious reasons why corporations, organisations and smaller companies are sometimes economical with the truth. They want to make their products sound as good as possible, and to limit any damaging revelations about themselves in order to protect their public image. They want to believe they are all good people even when their behaviour suggests otherwise. Sometimes they want to project a confidence in future numbers that they don't really possess, in the hope that this will help to make their predictions become reality.

We saw in the previous chapter how people are prone to interpreting information according to their existing beliefs. When presented with a piece of statistical information, we are more likely to doubt its veracity or relevance if it presents a picture that we are predisposed not to believe in. Obviously this can be one of the reasons why the employees or senior staff of an organisation will present false or misleading information. And in some cases they will know they are being disingenuous but persuade themselves that they are doing it for the right reason.

However, there is a tipping point beyond which it is not possible to make even psychological excuses. Some of the numerical deceptions in this chapter are too egregious to be regarded as normal, flawed human behaviour and instead veer towards the psychopathic or criminal (and note that those two terms are not synonymous ... there are plenty of psychopathic ways of behaving that aren't actually criminal!).

The main thing to remember as a customer, investor, analyst or employee is that organisations do lie, for all kinds of reasons, some better than others. But numbers can only be manipulated so far, so they are still one of our best guides as to the gap between what an organisation claims is happening and what is actually happening.

In that respect, real numbers are the best lie detector we have.

But as we'll see in the next chapter, the actual numbers and how the numbers can be made to look isn't always the same thing.

Bullet Point: Companies and organisations lie, and often misuse data; but real numbers are your friend and can help you see through the lies.

CHAPTER 4

Visual Distortions

THE STATISTICAL SMOKESCREEN

If you have a statistic that isn't helpful to your personal agenda, and you happen to be a liar, then you may find it helpful to resort to one of the visual methods described in this chapter, all of which help to introduce distortions or confusion when they are used to display numbers. Just remember that a badly (or cunningly) designed graphic can be as effect-ive as a smokescreen when it comes to hiding the truth.

TOO MANY AXES TO GRIND

The first thing to think about when it comes to statistical depictions of data is whether or not it is necessary to use a visual in the first place. A simple numerical table of data will often be unambiguous and clear, whereas many ways of displaying data immediately introduce potential distortions.

Secondly, it's important to make sure that the labelling of a visual doesn't introduce bias into the situation. Consider the bar chart below, which shows the number of sales (in thousands) of a brand of sunglasses over an eight-month period.

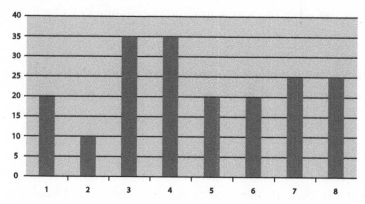

Now look at it again, but with a label.

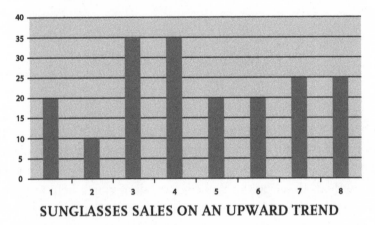

SUNGLASSES SALES ON AN UPWARD TREND

Immediately, the viewer is prone to looking at this chart in a different way. They might be prone to accepting that there really is an upward trend, ignoring the two months with the highest sales, or they may be sceptical about the intentions of the person displaying the chart. Or else they might note that eight months of a highly seasonal item is unlikely to give definitive proof of a trend. But whatever thoughts the label provokes, the mere use of it has filtered into the consciousness of the viewer and the chart is seen differently as a result. A more useful label would be something plainer like:

Sales of *** brand sunglasses April–December 2019

So, from the start, the way a graphic is presented can be a distorting factor.

Bullet Point: If a graph is presented with a big bold title, do your best to ignore it and focus on the actual information.

LIES, DAMNED LIES AND PIE CHARTS
Edward Tufte is an American professor of political science, statistics and computer science at Yale University. His book *The Visual Display of*

Quantitative Information was an early classic on the subject of how to display data clearly (and thus how to avoid misleading displays). He was trenchant on the subject of pie charts:

> Tables are preferable to graphics for many small data sets. A table is nearly always better than a dumb pie chart; the only thing worse than a pie chart is several of them, for then the viewer is asked to compare quantities located in spatial disarray both within and between pies – Given their low data-density and failure to order numbers along a visual dimension, pie charts should never be used.

Essentially the problem is that pie charts don't display information in a way that is easy to absorb. Consider this pie chart. Can you accurately rank the sections in order of size?

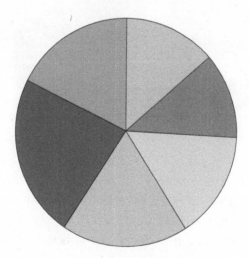

The answer is that from top right the segments show 15 per cent, 13 per cent, 17 per cent, 20 per cent, 25 per cent and 20 per cent of the total area (going clockwise). Here is the same information arranged in a bar chart: even though the bars are not in order of size, it is still much easier to compare and contrast them.

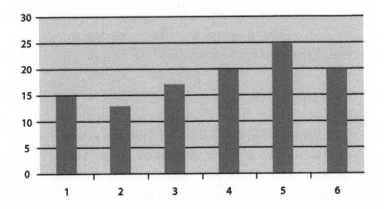

The only thing more duplicitous than an ordinary pie chart is a 3D pie chart. I'll talk more about the evils of 3D data displays shortly, but for now, take a look at this monstrosity.

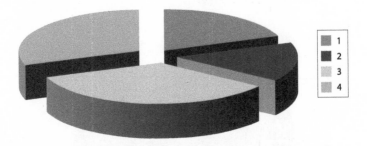

The numbers used to generate these segments were (from the bottom, clockwise) 35 per cent, 35 per cent, 20 per cent, 10 per cent. The problem is that the foreshortening of the 3D effect makes any segment placed at the front look much bigger than the same-sized segment placed further back, while the orientation of the slices affects how we perceive the surface area of the top of the 'pie', which is the only relevant part of the diagram: as a result, the 10 per cent slice looks considerably less than half as big as the 20 per cent slice, to my eyes at least.

Bullet Point: If you want to sneakily eat the biggest piece of a birthday cake, make sure your slice is on the opposite side of the cake stand to the friends or family members you are short-changing.

POWERPOINT, EXCEL, PAINT . . .

I have (and use) each of these programs. They all have their virtues (as does any other basic drawing, spreadsheet or presentation program). But they need to be used with caution. As I've mentioned earlier, when making images for PowerPoint, there is a temptation to oversimplify. Excel and other spreadsheet programs have options for creating charts and unfortunately include some of the most dubious methods of displaying information (such as 3D pie charts) alongside more reputable methods.

And when it comes to drawing programs, many programs make it easy to use dubious things like pictograms instead of bar charts: a pictogram is a chart in which images are used to represent the things being recorded in the data on display. For instance, here is a chart of the average pie charts used in global PowerPoint presentations over a five-year period.

What are we to make of this? It could be taken in at least two ways. Firstly, measured vertically, it suggests that the number of pie charts has gone from about one to about three. Horizontal background lines might clarify this if it is the intention. But the problem is that because a 2D image has been used, the right-hand image looks much more than three times as big as the left-hand image: in fact, the actual difference has been squared. In terms of area we have gone from 1 x 1 to 3 x 3 = 9.

A 9x increase could have been depicted thus:

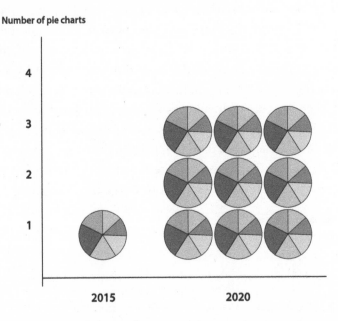

Alternatively, you could represent a genuine growth of one pie chart to nine pie charts by piling up nine little charts on top of each other in the right-hand column. But why do any of these when this entire method is subject to confusion. It would be so much better just to use a simple bar chart in order to avoid any confusion.

Here's another (slightly exaggerated) way that pictograms can go wrong.

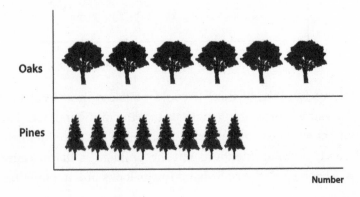

There's an old statistical cliché that you should never compare apples with oranges (or sometimes it is apples and pears). In this case, the problem is that the two images have different dimensions. There are actually more pines than oaks, but because they are thin, the visual impression is the opposite. For another example, imagine comparing bananas with apples in a pictogram that uses images of the fruit. How would you make it unequivocal how the two were to be mathematically compared? There are numerous different ways in which the pictogram could be confusing and unclear and pretty much no way it could be done with clarity.

Bullet Point: If in doubt, don't use pictograms. Even if you're not in doubt, don't use pictograms.

The Horrors of 3D

If the mistake introduced by a 2D pictogram is to square the difference, then the difference introduced by a 3D version can theoretically be as much as to cube the difference. Here's a misleading graphic showing the growth in sales of caravans over a five-year period.

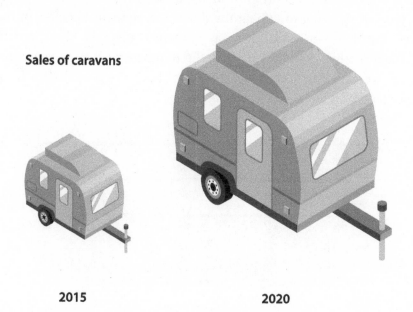

Sales of caravans

2015   **2020**

If this is intended to show a threefold growth in sales it is entirely misleading. The image on the right is three times bigger in every dimension and so it looks twenty-seven times as big. But if it is intended to depict a twenty-seven-fold growth, it doesn't really do the job either. And on the page, it is taking up about nine times as much space, so it is about as ambiguous as you can possibly be with such a simple graphic.

Even a relatively sober 3D bar chart is dangerous. For instance, here is a chart for the number of completely pointless 3D bar charts used at a given company over a three-year period.

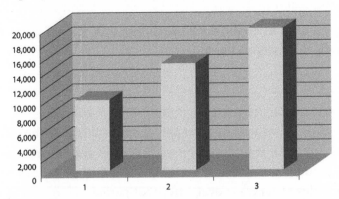

Now, because the bars only grow in one dimension, they don't make the mistake of squaring or cubing the difference between the sizes of the bars. But there are numerous ways in which this falls short. Firstly, consider the horizontal orientation: the numbers used to generate this image were 10,000, 15,000 and 20,000. If I add horizontal lines from those points on the X axis, I get this image.

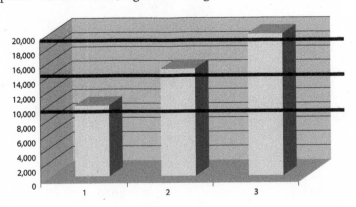

I'm not clear what heuristic was used to generate the position of the columns, but they don't seem to line up with the lines on the back wall or with a line projected across from the X axis. The explanation is presumably that, because the bars are located neither at the front nor at the back of the little box they are contained within, we have to use perspective lines to understand what is going on.

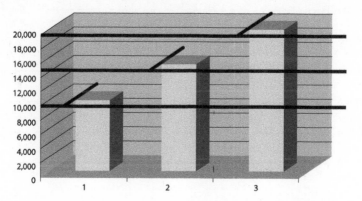

However, that is not an easy calculation to make without actually using technical drawing skills to draw in the lines, so the image is extremely misleading when it comes to the actual height of the columns. This could be slightly improved on if the front edge of the bars reached the top of the space, and the back edge reached the back of the space, but it would still be ambiguous and hard to read. The only real way to understand this would be to refer to the actual data, which entirely defeats the point of using a visual display in the first place.

Bullet Point: If you want the opportunity to make a deeply misleading graphic representation of some data, 3D is your friend.

Labelling Your Axes

If you ever see a visual chart in which the axes aren't clearly labelled then you should immediately feel your hackles rising. There is no good way to understand a graph if it isn't entirely clear what the axes actually represent.

However, that isn't the only way that axes can be used misleadingly: there are situations in which it makes sense to use a logarithmic scale in a graph rather than a linear scale. Here is a graph that shows a price that doubles every year for a period of twenty years.

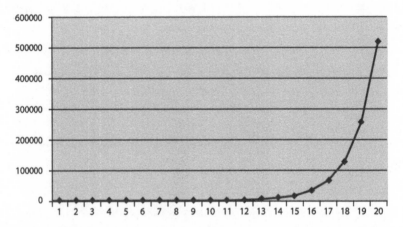

Exponential growth is a scary thing in many contexts: when it comes to the use of natural resources, for instance, it can lead to a resource heading for complete depletion. But when it comes to something like growth in the amount of data that can be handled by a computer chip, it is not so terrifying. But it can be helpful to find different ways to depict this kind of information. So we might choose to use a logarithmic scale. In the image below, the new bold line shows the same information as the graph above (I've left the original line in for comparison).

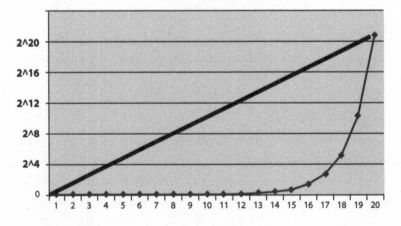

The crucial thing here is to be explicit about what kind of scale is being used. Even if you are talking to a scientific or mathematical audience, who will be more familiar with the concept of a logarithmic scale, it is important to be absolutely clear in the labelling and description of such a graph.

Bullet Point: The axes of a graph should always be clearly labelled and the scale should be obvious.

CONFLICTING AND TRUNCATED AXES

Another thing to look for in any graph is whether or not tricks are being played with the axes. Here is a graph showing interest rates from 2008: note that they appear to be increasing at a terrifying rate.

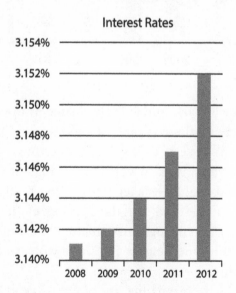

The trick here is the truncated Y axis. A more honest version of the same graph, with the Y axis starting at zero would look like this:

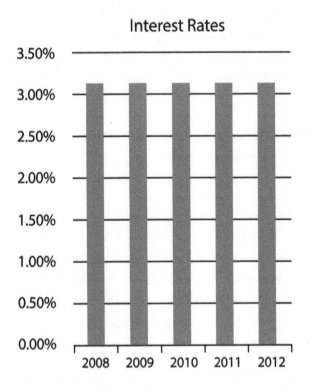

The difference is stark, and entirely created by the manipulation of the Y axis. In general there is rarely a good reason for a graph where the axes don't start at zero, as it will always create a distortion in the gradientand any trend. It's also worth noting that the first graph above would have looked slightly less terrifying if the X axis had been more widely spaced, so also look for the spacing of the graph. This is another point at which pictograms can blur a message, since they can be of varying widths and heights and these will also affect any gradients in the graph.

A related bit of trickery is the use of mixed axes. Suppose the evil human resources director puts the graph below in front of the board in order to argue that staff cuts have made no difference to the company's sales performance, so there is a good case for another round of redundancies:

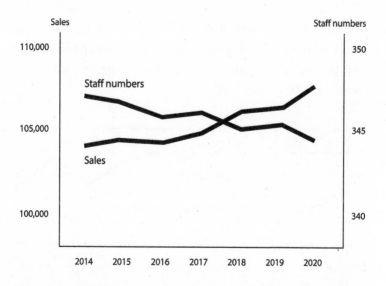

Bullet Point: A graph should have clearly labelled axes and if they are truncated or mixed then the graph is probably meaningless.

The first thing to note is the truncated axes on both sides. This makes it very hard to judge whether or not the change in the two numbers is significant or not. (A Y axis starting at zero would show both measurements looking more or less flat.) The second thing is that the two Y axes don't in any way match up to one another: if they were extended downwards towards zero, they would reach zero at wildly different points, which means that they are entirely invalid as the basis for comparisons.

So, remember that one of the key giveaways that there is something fishy about a graph is if it is presented without labelled axes. If you don't even have the information presented that allows you to untangle the confused over-truncated or mixed axes, then you are completely in the dark about what information the graph is actually conveying.

MISSING DATA
This time, let's start with the honest graph and see how we might be able to manipulate it. Let's say I want to impress my bank manager in a

meeting by showing them that my income as a freelancer has been increasing over the last ten years and is still on an upward trajectory. So I get out the old tax statements and draw up the following graph with income in £1,000s on the Y axis.

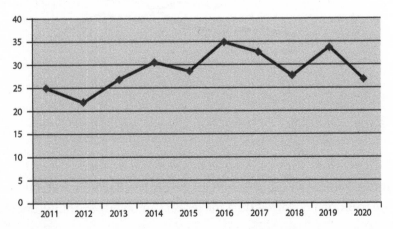

It doesn't look great, does it? The final position is close to the starting one, and there is a lot of volatility. I could draw a scatter chart with a misleading trend line (see below), but that might be a bit obvious. Perhaps instead I could miss out every other year, show it to the bank manager quickly and hope they aren't concentrating enough to notice that the data is incomplete . . . Here is the same graph using every other year.

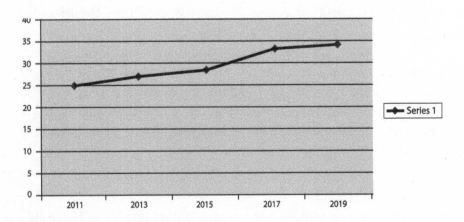

Bingo! This looks much better.

As a disclaimer, I should mention that the income in the graph above is entirely made up and not based on my actual earnings. And even more ludicrously, I've implied I might be able to have a meeting with my bank manager, which is clearly not the kind of thing that happens in the twenty-first century: the chances are I would be ticking boxes on an internet form or speaking to a faceless voice down a telephone.

But the basic trick of omitting data to form a better picture is one that can be transferred into many other situations from sales charts to climate change analyses or whatever.

Bullet Point: Missing data is a red flag that shouldn't be ignored.

SCATTER CHARTS

It's always a good idea to keep a sceptical eye on any trend lines that are added to a scatter chart. Consider this set of points on an X–Y coordinate graph. It represents the number of bird species spotted at a reservoir in the last twenty-three years. Does it show an upward trend, or a downward trend?

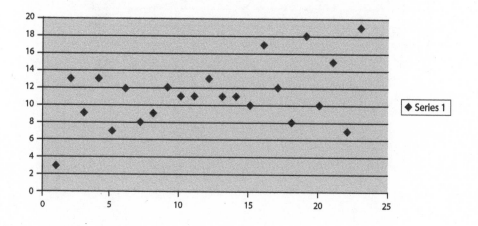

Well, if you are a true optimist, you might draw in this trend line:

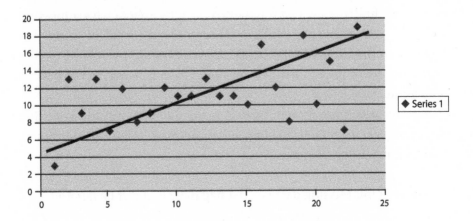

On the other hand, a pessimist can also find a suitable trend line:

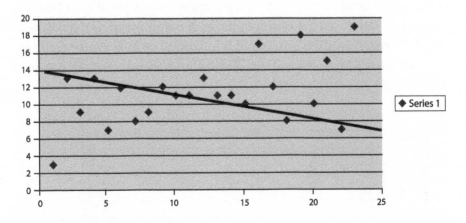

And there is always the 'glass half full' option, which falls somewhere in between the two extremes:

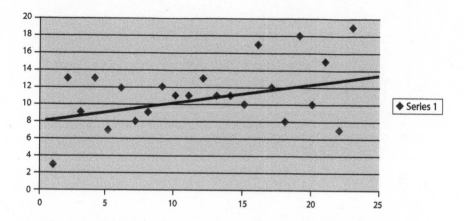

The point is that all of these trend lines look at least slightly persuasive. The chances are that to make more sense of this data we might need to compare it against other significant factors: for instance, we might try plotting it against the severity of the winters or similar climate-related factors in order to investigate its significance.

The way I actually constructed the set of points in the first place was to create an 'upward trend' set of points and then to intersperse this evenly with a 'downward trend' set of points. Not all scatter charts will be as ambiguous as this, but even on clearer trends, it is perfectly possible to exaggerate the significance of a trend.

For an example, look at this scatter chart of annual sales (in millions) at a company. This chart has a clearer upward trend.

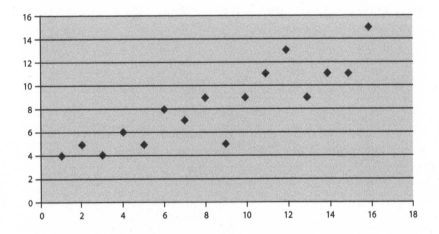

When the sales director, who is always the most ebullient member of the board when it comes to forecasts, adds a trend line, he opts for this one.

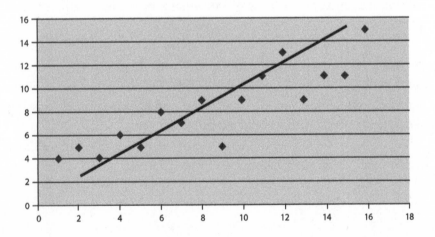

However, the finance director is a more cautious soul, so he sees a rather different pattern.

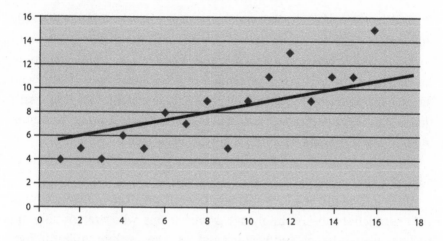

Bear in mind that the human mind looks for patterns, and readily accepts them if you give it a nudge in the right direction. Any of the trend lines above could be accepted as a genuine attempt to identify a trend, but any of them could also be used deliberately to exaggerate or underplay the significance or direction of a trend.

Bullet Point: Be very sceptical about trend lines on scatter graphs.

REGRESSION ANALYSIS

Incidentally, when statisticians draw trend lines on graphs, they do use more formal methods than the previous section suggests: in particular, they use regression analysis (or even multiple regression analysis), which is essentially a collective term for various analytical methods that allow us to estimate what the average or average trend line is in a set of data that shows some volatility around that average or trend.

This is, in theory, a respectable mathematical method. But note that any type of regression analysis is subject to potential flaws. You need to make assumptions about any dependent or independent variables that might affect the data, about potential error terms and whether any unknown factors may be significant. And then you need to choose which method of regression analysis you use, which means that there is scope for a cunning analyst to choose the method that produces the trend they want.

And, at the end of the day, it doesn't matter how cleverly or honestly the analysis has been carried out if you don't treat the results with common sense. Imagine a scientist goes to a lot of trouble and a lot of research and produces some undeniable evidence that sports teams who bump fists regularly are more likely to win their matches. He might take this to a gullible sports coach who will insist that their team keeps bumping fists at every chance they get through the season.

But still the team has an awful season, with no victories at all. What's gone wrong?

Well, maybe this is a case that takes us right back to Statistics Lesson 101: correlation doesn't imply causation. Teams who are winning are perhaps feeling happy and positive and this causes them to bump fists rather than the other way round.

The same might apply to a study that conclusively revealed that people driving Audis have more accidents per mile than those who drive Renaults and thus are less safe, without asking the question of whether or not the two cars attract different character types (for instance macho, competitive types as opposed to cautious, risk-averse types), which might be the underlying cause of the correlation.

The point being that it doesn't matter how subtle the maths is or how big the computer is that manipulated a set of data. What matters is treating the output with common sense.

Bullet Point: If someone is trying to blind you with science or with regression analysis, don't let it overwhelm your judgement.

SIMPSON'S PARADOX

While we are on the subject of trend lines, there are numerous other reasons to be careful about their use. Just to give one example, suppose that the head of a school wanted to advise his pupils that the more time they spent studying, the better their results would be. He samples sixteen students at random, checks how long they spend studying a particular subject and how well they did in the exams, and obtains the following scatter graph:

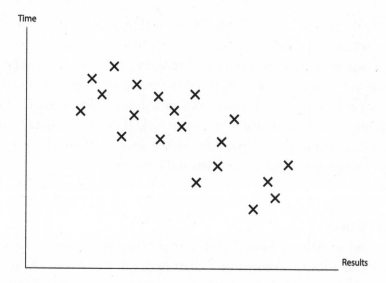

This looks like it proves his hypothesis wrong: clearly the trend line here would be down from left to right, thus proving that the less time a student studies for, the better his results. Luckily, the maths teacher is on hand to smugly point out the problem. Let's divide this set of data up into different subject areas: the circled areas separate the data out into students who were tested and asked about study time for English, Science, Maths, History and Humanities:

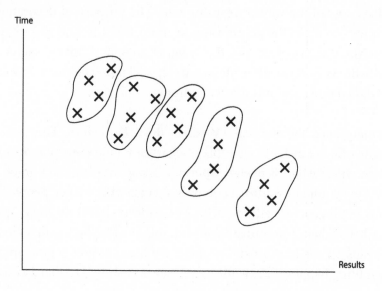

Clearly there is indeed an upward trend within each subject, even if the exact relationship and gradient varies slightly. This is an example of Simpson's paradox, in which a trend appears in several different groups of data but disappears or reverses when these groups are combined. It's another reason why it is best to avoid combining different kinds of data on a single chart, and also a reason to look at any set of data from as many different angles as possible, as the illusion of trends from one angle may disappear when you view it differently.

CHARTJUNK

The great statistician Edward Tufte (see p. 110) used the term 'chartjunk' to refer to a visual element in a chart or graph that doesn't help us to comprehend the data being displayed and which might thus distract the viewer from clearly comprehending the data. Any kind of marking, picture or colour can be referred to as chartjunk, as can overly empha-sised grid lines, wacky fonts, pointless text, shading or gradients or surrounding decorations. Basically, a graph or chart should be presented in the most stripped-down way possible. Tufte said:

> ... the interior decoration of graphics generates a lot of ink that does not tell the viewer anything new. The purpose of decoration varies – to make the graphic appear more scientific and precise, to enliven the display, to give the designer an opportunity to exercise artistic skills. Regardless of its cause, it is all non-data-ink or redun-dant data-ink, and it is often chartjunk.

I've mentioned the satirical TV series *Brass Eye* before. The 'Animals' episode offers a brilliant example of chartjunk (and generally meaning-less graphics) when it flashes up a needlessly overdesigned graph that plots the 'number of animals abused against what makes people cruel versus intelligence of either party'. The pattern shown on the graphs is described as being 'so unreadable, you might as well draw in a chain of fox heads on sticks'. And at this point, the narrator points out, 'an inter-esting thing happens: the word "cruel" starts flashing'. It's hard to beat as an example of the overly sensationalised graphics that can be used in political broadcasts or the worst kinds of television news.

Bullet Point: Beware chartjunk.

GETTING THE MEASURE OF THE LIES

There are a few ways of measuring the distortion of a graphic statistics display. Firstly we can measure the **lie factor**. If, for instance, a graph has a truncated Y axis it might make the difference in a quality look as though it is a 50 per cent increase, when it is only really a 10 per cent change. The lie factor is calculated by dividing the size of the effect implied by the graphic by the actual effect. If the result is 1, the graph is accurate. If it is higher than 1, the graph is exaggerating the effect. So in the example above, the high lie factor of 5 means that this is a very dishonest graphic. By contrast, a lie factor that is lower than 1 is masking or concealing the severity of an effect.

Another useful measure is the **data-ink ratio**. This is calculated by dividing the amount of ink (or pixels) used to display the data, and the amount used in the graphic as a whole. The data-ink ratio should be as high as possible, otherwise the other elements in the graphic are in danger of overwhelming the data. A similar measure is **data density**, which is calculated by dividing the number of entries in the data set used in the graphic by the area of the graphic. This is more of a rule of thumb than a clear indicator, but it is always worth looking at a graphic sceptically if the amount of actual data presented is very small compared to the size of the graphic.

A final consideration is **lexicographic cost**. The amount of words used in a graphic should be enough to clearly explain the axes, and what is being measured, but anything beyond that will simply be a distraction.

Bullet Point: Honest graphics aren't cluttered, overly decorated or full of words.

VENN DIAGRAMS FOR NERDS

I was once in a publishing meeting where one of the editors was presenting the stats that justified their proposed purchase of a book: it was an Instagram star's book about their hobby, fly fishing. They pulled up the social media stats, followers, likes and so on for the Instagram star, then pointed out how many people are members of the fly fishing community, and finally added the two figures together to suggest that there was a huge market for the book.

In moments like this, I used to be fond of irritating my colleagues by drawing Venn diagrams (which may explain why I work at home on my own these days). Anyhow, the diagram for this would look something like this:

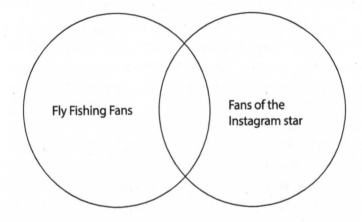

The danger in a situation like this was that the editor was assuming that the market for the book could logically be expressed as:

'Fly Fishing Fans **OR** Fans of the Instagram Star'

Where it was perhaps more likely that it would be:

'Fly Fishing Fans **AND** Fans of the Instagram Star'

The former would include the entire area of the Venn diagram. The latter would only include the intersection between the two. This is actually a case where a simple graphic may have helped focus attention on the conundrum. But it is also a good idea to treat Venn diagrams with caution, since they are so widely misused. This is especially true when it comes to social media memes; you'll often see comical versions of this kind of thing:

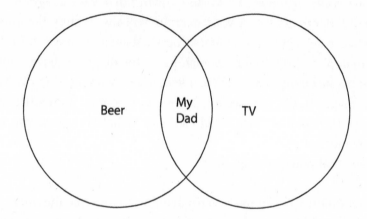

OK, to be a viral meme, it'd need to be a lot funnier, to have more colour and chartjunk and so on, but you get the idea. The point is that this isn't how Venn diagrams work. Your dad isn't a type of beer, nor is he a TV. This would be more correct:

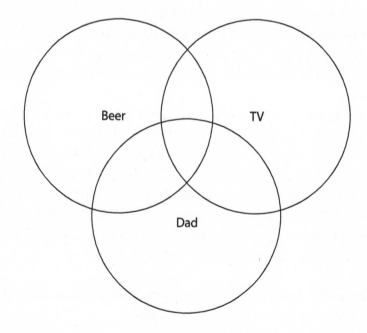

Of course, the intersecting areas in this diagram would be empty because no one is both a dad and a beer, a beer and a TV or a dad and a TV (unless you mean a transvestite, in which case there would be a few in there . . .)

To get really pedantic, it's worth knowing that Venn diagrams are a subset of Euler diagrams. A Venn diagram will always show the intersection between sets, even if the intersection is itself an empty set. An Euler diagram depicts the actual relationships, so the diagram above would be three separate circles (with the TV one marked 'Television' for clarity). In a Venn diagram, the intersection between two circles shows the things that have the qualities of both circles, not 'What you get if you mash these two ideas up'.

I hope that makes things clear.

> **Bullet Point: You can use a Venn diagram to describe the inter-section and difference between two or more sets. Or you can use it to make a joke about kittens on Instagram, but don't expect any maths pedants you might be friends with to be laughing . . .**

CONCLUSION: VISUAL DISTORTIONS

A good graphic representation of a data set can be invaluable. Just as a picture conveys a thousand words, a good graphic will boil down the important information and present it in a way that is intuitive and easy to grasp.

However, there are too many situations in which people's main motivation is to conceal, distort or exaggerate the truth. So you need to be on your guard any time someone who has a vested interest is showing you a graphic. It is far too easy to get fooled by dodgy axes, chartjunk, dubious trend lines and any one of the other many pitfalls that graphics can introduce. It's always worth asking for the actual data set if you have any grounds for suspicion, and it's always worth studying a graphic carefully to see if you can spot any sleight of hand.

> **Bullet Point: A graphic can make a data set easier to understand. But it can also make a lie easier to swallow.**

CHAPTER 5

Retail Trickery

....

WHERE'S THE EXTRA?

When you are shopping, it's worth being very aware of the many ways in which retailers and product manufacturers can twist the truth. For instance, I recently went to the supermarket to buy a very specific brand of shampoo: I'm a bit of a hapless shopper, and having previously bought the 'glycogel' variety rather than the 'protein + pearl' variety by mistake, my wife suggested that this time I avoid repeating the error by taking the empty bottle I was supposed to be replacing.

Having scanned the shelves at some length, I isolated a bottle that had all the right words on it in the right order, but there was one puzzling difference. The full bottle on the shelf bore an extra flash claiming that it had '25 per cent extra free'. This was in spite of the fact that the two bottles were exactly the same size and, for extra confirmation, both bottles said they contained 500ml.

I eventually spotted a vanishingly small asterisk after the 'extra' claim; and in the middle of the back label, in even smaller print, there was an explanation:

* Compared to our standard 400ml-size bottle.

Now, because I'm that sort of person, I checked the shelves in that supermarket for a 400ml bottle, without luck. I also checked all the other local supermarkets and pharmacies, and it didn't seem that anyone actually stocked the 400ml bottle. It is true that I found it on the manufacturer's website, so I was forced to give them the benefit of the doubt and accept it did exist rather than, for instance, making the unjustified assumption that they were just pretending it did in order to justify the 'extra' claim. But I still think it would be a bit of a stretch to claim that the bottle size that no national chains stock is your 'standard' size.

Of course, this sort of thing is standard practice. I used to know a package designer who had been taught some of the tricks of the trade at college. He told me about his lecturer showing them a can of beer that had a 'genuine' 10 per cent extra claim compared to the usual size of can. The catch was that the '10 per cent extra free' claim was printed on a banner running around the can that actually took up about 20 per cent of the height of the can, thus visually implying that the free bit was bigger than it really was, in spite of the claim being technically true.

Bullet Point: Is the extra bit really free? Is it even extra?

Multi-buy Bargains

Another common problem relates to multi-buy deals in supermarkets: one investigation by *Which?* magazine looked at thousands of online prices at the main supermarkets. For around 10 per cent of the products it checked that were advertised as multi-buys (for instance '£9.99 for three'), the multi-buy price was actually higher than buying the items individually. One supermarket was selling individual pizzas of a particular brand at £1. However, when the pizza was put into a multi-buy deal at £4.50 for two, the 'price' of an individual pizza was advertised as being £2.50. Packets of sweets that ordinarily sold for £0.34 were being sold at £3 for four, meaning that the 'deal' cost shoppers an extra £1.64. At a more trivial level, four-packs of soup were being offered for £2.99 where the individual tins that were sitting on the same shelf were £0.75.

So, it's only a penny, and not much of a 'deal', though I suppose, as one of our leading supermarkets is fond of pointing out: Every Little Helps.

Bullet Point: Multi-buy deals and offers aren't always the bargain they claim to be.

SHRINKFLATION

Let's say you are a manufacturer who wants to protect or increase your margins. One way to do this would be to put your prices up. But then canny shoppers would be quick to spot the increased price. So the clever trick adopted by many manufacturers is simply to make the product cheaper to manufacture. You can sometimes do this by using cheaper ingredients but you can also simply make your product smaller.

In the UK, the Office for National Statistics recently issued a report on the effects of shrinkflation. In the two-year period up to mid-2017, they found 206 products that had decreased in size (as against half as many that had increased in size).

Some manufacturers blamed this on the outcome of the Brexit referendum and the ensuing drop in the value of sterling. For instance, Birds Eye explained the fact that they were putting fewer fish fingers into a pack thus: 'The cost of many of our raw materials has risen since the EU referendum.'

Other examples included Mars cutting the size of individual Maltesers and M&M packets by up to 15 per cent, McVitie's cutting the number of Jaffa Cakes in a pack from twelve to ten, Tropicana cutting the size of the cartons for its juices, and Doritos cutting the weight of tortilla chips per packet.

One of the most comical moments came from Toblerone. If you don't know the brand, it is a triangular nougat chocolate bar: individual triangles of chocolate project up from a long rectangular base. Rather than make the whole thing shorter and including fewer triangles per bar, Toblerone simple made the triangles slimmer, leaving a considerably bigger space in between each triangle.

Social media instantly latched on to this, as floods of comparison photos showed the reduced state of the new bars: the manufacturer, Mondelēz International, eventually gave up and made the bar in its original shape, but bigger.

Of course, it then came with a higher price.

For me, the final insult with shrinkflation came with a brand of washing powder I use. The original packet had gradually shrunk to the point where you only got a handful of washes out of each box. Then one day I visited the supermarket and found a box marginally bigger than the original size labelled as a 'New Family Size' pack . . . at double the original price!

Bullet Point: If you think your chocolate bars keep getting smaller, it's not just because you are getting old and crotchety.

THE SHOPPERS' FRIEND

Retailers, and supermarkets in general, like to project the image of being the shoppers' friend, but of course, at heart they are businesses, whose main target is to increase the bottom line. It's worth a brief look at the recent history of UK supermarkets' interactions with dairy producers to see some of the pressures this creates.

Firstly, in 2007, some of the major supermarkets were fined £116 million by the Office for Fair Trading, after admitting that they had been fixing the price of milk, cheese and butter: it was estimated that the practice had cost consumers about £270 million.

In previous years, the supermarkets had been in breach of the 1998 Competition Act, which aims to stop businesses from colluding in an anti-competitive way. As a result, customers were getting charged 3p extra for a pint of milk, and 15p extra for a quarter-pound of butter, or for a half-pound of cheese.

So that's one danger when it comes to protecting the consumer. (Incidentally, price fixing can work the other way round. In 2019, Walmart in the US filed a lawsuit in Arkansas claiming that a group of US poultry companies had been conspiring to inflate chicken prices between 2008 and 2016.)

Fast-forward a few years and a different problem was affecting the dairy industry. In 2015, dairy farmers called for a boycott of some of the leading supermarkets because the price they were insisting on paying for milk and other dairy products was making farming those products uneconomical. At the time the supermarkets were engaged in a price war that led them to cut the price of basics such as a pint of milk. Rather than taking the whole hit for this, they were allegedly imposing an unsustainably low price on the producers.

This is an example of 'buyer power', which is the ability of certain types of powerful buyers like supermarkets to obtain favourable terms, more so than would be available in a more competitive marketplace. The supermarkets buy large volumes and while there are some food producers large enough to have significant bargaining power, the supply side of the industry is mostly made up of smaller players who don't have that bargaining power. (A similar issue tends to affect the book trade where large players like WH Smith, Barnes & Noble, Waterstones and Amazon have a great deal of buying power.)

When you see BOGOFs and WIGIGs ('buy one get one free' and 'when it's gone it's gone') promotions, you may feel the supermarket is doing you a favour. It's always worth bearing in mind that these tend to be negotiated in advance with the suppliers, who will feel under a great deal of pressure to cut their prices to the bone in order to get the large-volume orders that result. Supermarkets often make up the bulk of the national grocery market. For instance, in Austria, the top three supermarkets make up 82 per cent of the market, while in Finland it is an eye-watering 88 per cent. Obviously this gives them a huge advantage when it comes to small suppliers, who will struggle to find much of a market beyond them.

In some jurisdictions there is legislation about the abuse of buyer power. This can include things like charging fees to suppliers to be on an approved list, forcing suppliers to take the risk on new products, charging for shelf space, demanding unexpected or retrospective fees from suppliers, insisting on returning unsold stock (even fresh produce that can't be resold) and late payments.

Another significant area of worry for suppliers struggling to survive is the use of own-brand items. This is an area of retailing that has grown hugely over the last couple of decades. In Australia, the share of supermarket food sales made up of retailers' own brands has recently been estimated at 25 per cent. Given that shelf space is limited, each time the retailer expands their own-brand range into new products it is likely that a supplier is the one losing out.

Incidentally, you might want to take a moment to examine the labels on own-brand products now that supermarkets have expanded beyond the basics ranges. A recent news story examined the difference between the cut-price version of products and the own-brand version marketed as a more upmarket product. In numerous cases they found that products that were the exact same weight and contained the exact same ingredients were priced as much as twice as much for the non-basics range. Often the only difference is that the 'luxury' version has a few more adjectives in the description and a better-designed label than the 'basics' version.

Another trick to watch out for: in the health-conscious age we live in, it is now highly lucrative for the supermarkets to provide products that are, for instance, gluten-free or dairy-free. When they sell a product in their 'Free From' range, it may be marked up to a much higher price than a virtually identical product elsewhere in the store. For instance, in 2016, Tesco in the UK was selling a gluten-free tomato ketchup in its free-from range at £1.20 while elsewhere in the store the standard Tesco's Tomato Ketchup (also gluten-free) was only £0.65.

Bullet Point: Supermarkets and suppliers are businesses, so think about what might have happened behind the scenes of a cut-price offer.

RETAIL TRICKS

Here's a quick rundown of some common ploys that help to inflate a company's profits at our expense:

- Popcorn in the cinema: there are plenty of overpriced products on offer, but one of the most extreme mark-ups around is that carton of popcorn in the cinema. Often sold at about £4 or $6, the mark-up can be well over 1,000 per cent. The actual popcorn only costs about a dime or less (about £0.07), and, gallingly, even the box costs the cinema more than the contents.
- While we're on the subject of remarkable mark-ups, beware the hotel minibar. One New York hotel was recently charging a 1,300 per cent mark-up for a pack of gummy bears.

The New £9.99

Retailers have always used the old trick of pricing items at £9.99 instead of £10. The idea is to play on our cognitive biases: even though the price difference is only £0.01, £9 sounds cheaper than £10, and we like to think we are getting a bit of change from a tenner.

You may have noticed a new variation on this: I believe it started in hipster bars and cafés, possibly as an affectation in the first instance. This is when prices are given in a single number: for instance, rather than Avocados on Toast being listed as costing £7.00, and Coffee at £4.00, the brunch menu will say something along the lines of:

Sustainably Sourced Avocados on Hand-Toasted Sourdough Flatbread	7
Flat White with Organic Almond Milk and Crumbed Cinnamon	4

Again, there is a cognitive bias operating here. Rather than thinking about 700 pennies, we are being invited to think of 7 pounds: now that a pound has become a fairly low-value unit due to inflation, this can seem like a smaller amount than £7.00 or even £6.99.

- The zero-interest purchase: my wife recently bought a sofa for about £800. She was pleased to find that she could spread the payments out over a year, and pay zero interest. So she ticked that box and worked her way through the website, only to find on the payment page that the zero-interest offer was only available if you paid an additional £140 charge upfront. This meant that the zero-interest offer was actually incurring 17.5 per cent extra costs. Interest, in other words. (She changed tack and chose not to take advantage of that great offer, since she could do it on a credit card at a far lower rate.)

- In a similar vein, be wary of zero-interest credit cards with balance transfer offers: the rate that the debt will eventually revert to is often very unfavourable.
- Watch out for the extended warranty on high-price products: as I'll discuss in a later chapter (see p. 222) these are usually wildly overpriced forms of insurance, carefully designed to protect the product only for the period when it is least likely to actually malfunction. Similarly, watch out for any add-ons you are offered at the point of sale. Shoe shops often try to sell products such as suede protectors when you buy a pair of suede shoes: the same product is almost always far cheaper to buy elsewhere, if you really need it in the first place.
- The law on sales items varies in different countries, but in the UK a sale item need only have been on offer at the old, higher price for twenty-eight consecutive days within the last six months. So retailers can hike up the price in the run-up to a sale, and then 'reduce' it back to the original level (or close to it). As an additional insult, the increased price may only be available in a few remote outlets that are unlikely to make many sales in that twenty-eight-day period anyhow. One recent investigation by a shopping comparison site found that major kitchen suppliers had increased their kitchen prices by up to 130 per cent in the month before Christmas (when few people buy kitchens) in order to be able to 'slash the prices' for the January sales.
- Women may want to examine beauty products carefully before purchase: while the smell and packaging will vary, there can otherwise be little difference from the same products marketed for men, other than pricing. One razor brand recently marketed a pink version of their razor for £7.00 while the men's identical (albeit less colourful) version was £1.30 cheaper. The same applies to shaving gel marketed for women, for instance. And when it comes to anti-ageing creams, it seems as though companies have cottoned on to the fact that consumers believe a higher price will guarantee miraculous results, so the price of

something that is basically a jumped-up moisturiser can be astounding.

- Always bear in mind the cost of replacement parts: you've probably noticed that while you can buy a printer for a fairly cheap price, the replacement ink cartridges (which are specific to the brand or model) are hugely expensive (unless you can find a generic replacement). Similarly, buying a razor with a couple of blades can be much cheaper than going back to buy the replacement packs of razor blades.

- In the supermarket or any kind of shop, bear in mind that they are masters of psychology. Sales products are often displayed on the end of the aisles to give a false impression of how many bargains are available. Expensive objects are often placed at eye level while the cheaper ones are high up or low down (since we are most likely to buy the first one we spot). And overpriced impulse buys are placed by tills (as anyone with a toddler who loves sweets may have noticed).

- Yo-yo financing: this is a practice that is illegal in some states of America, but which you might want to be aware of. A car is sold with the promise of financing, and the customer is allowed to take the car away, then the dealer tells them that the financing has 'fallen through' and offers a new, higher-priced finance deal. Having made the emotional commitment to the car, it is all too easy to give in and accept the new deal, no matter how ruinous the rate might be.

- When it comes to drugs, bear in mind that the short patents available to drug companies mean they work hard to present their version of a generic drug as being superior in order to extend its shelf life. In my local chemist, paracetamol currently sells for £0.22 in an unbranded packet, but for up to £1.60 for the branded one, which is over seven times as much for the same actual chemicals.

Bullet Point: You're not paranoid: they are out to get you.

Partworks

If you're not familiar with the term, partworks are products sold in instalments: examples include magazines where each issue comes with a toy (such as a doll in national costume), which will build up week by week into a collection, and variations on the theme of 'build your own spaceship', where the parts of a toy you build are supplied week by week. There are a few potential problems with partworks. One comes when a manufacturer releases the first few issues, but then decides the line is unprofitable and withdraws it. But if they do last, they can also be extremely costly.

Manufacturers are obliged to be honest about the potential full cost, but when you see or hear a TV or radio advert, the costs will sometimes be mentioned only briefly in the smallest print possible. One recent offer from the company Eaglemoss (who are not the only company to offer similar deals) allowed the lucky reader to collect magazines along with parts for the construction of the iconic car from the movie *Back to the Future* with working lights. You could immediately buy part one for £1.99 and part two for £4.99, which seems fair enough. Until you realise that the standard price for an issue was £8.99 and there were 130 issues. That means the full working model would take two and a half years to collect and would cost £1,157.70.

As one newspaper noted at the time, it was possible at the time to hire a real DeLorean car for a week for £150 plus transportation costs, which would work out a lot cheaper and probably be a bit more fun.

CONCLUSION

Recent scientific research has shown that up to 29 per cent of 12,675 purchases in large retailers in seven countries were overpriced, misdescribed or otherwise misleadingly advertised.

OK, that was a lie: I was just testing you. You hopefully spotted it from the use of overly specific Potemkin numbers. But the point of this chapter is that there are many pitfalls when it comes to parting with your money, so always examine any deal or offer with a sceptical mindset, and with your calculator to hand.

Bullet Point: Look after the pennies, because if you don't someone might con you out of the pounds.

CHAPTER 6

When Scientists Lie

WHAT LEADS TO SCIENTIFIC MISCONDUCT?

For the most part we place a fairly high level of trust in scientists. Unlike politicians and advertisers, we know that the fundamental aim of most scientists is to uncover the truth rather than to sell us a lie. So there is good reason to be reasonably trusting.

Unfortunately there have been quite a lot of cases in which scientists have fabricated or distorted the data. Which leads to the question of what causes a scientist to lie.

The most obvious motivation is ambition and the pressure of a scientific career. Scientists need a good reputation in order to get funding and acclaim, and high-profile findings can be a huge boost to their career. In addition, it is all too easy to fabricate results in certain areas of science. Experimental results can be hard to assess or replicate, because noise, the specific tools used and external data can all affect the exact outcome of an experiment. In the end, if an experiment can't be replicated, it may be discredited but, as we'll see, this may not mean that the scientist entirely loses their reputation.

It may not be a coincidence that, at the same time as we are living in an era of fake news and motivated belief, there has been a surge in the amount of fake science. A 2012 study by the Proceedings of the National Academy of Sciences (PNAS) considered 2,047 papers that had been listed as retracted (meaning that they had been debunked and withdrawn) on PubMed, a database of biomedical and life sciences research. The researchers found that 67.4 per cent of retractions 'were attributable to misconduct'. Out of that figure, 43.4 per cent had been withdrawn due to 'fraud or suspected fraud', while 14.2 per cent were duplicate publications and 9.8 per cent were due to plagiarism. Just 21.3 per cent of the retractions had been made because of an error by the scientists who had put their name to the research.

More worryingly, they found that: 'The percentage of scientific articles retracted because of fraud has increased more than 10-fold since 1975.' There were only around 100 retractions based on fraud or suspected fraud in the two decades from 1977 to 1996. However, there had been 150 in the five years from 2002 to 2006, and then, from 2007 to 2011, over 400 papers had been withdrawn because of fraud.

The *Scientific American* blogger Ashutosh Jogalekar was discussing this study when he noted that 'Biomedical research is likely going to continue to invite fraud because of its sheer and growing complexity. In addition the monetary benefits, fame and visibility associated with potentially important results in fields like caloric restriction are immense, leading to even more opportunities for fraud and wishful thinking.' So it is a combination of ambition, vanity and greed that tends to lead to fraud.

This can be as simple as crude lying, as in the 1974 case of William Summerlin from the Sloan-Kettering Institute in New York, a leading biomedical research centre. Summerlin claimed that he had transplanted human corneas into rabbits. More bizarrely, he faked transplantation experiments in white mice by simply using a pen to blacken patches of their skin with a pen, an extraordinarily crude form of forgery.

But more often, scientific fraud relies on dodgy numbers: whether the data is invented or distorted, the reporting of untruthful numbers can be hugely significant. In this chapter we will look at a few examples of the consequences of some specific bits of scientific misconduct and pseudo-scientific gobbledegook.

(Incidentally, if you don't believe in man-made climate change or think that vaccination causes autism, you may not like this chapter. You could always skip it and move on to the next chapter, though I'd be offended if you did.)

The Anti-Vaxxer Guru

Andrew Wakefield was a British doctor and researcher who is widely regarded as having fuelled the anti-vaccination movement. This is hugely significant as the decline in take-up of the MMR vaccine in particular is leading to alarming new outbreaks of mumps, measles and rubella, putting many lives at risk as 'herd immunity' is weakened.

Wakefield's main interest was in inflammatory bowel disease: he trained as a gastrointestinal surgeon. He claimed that he was approached by the parents of an autistic child who was suffering with gastrointestinal problems in 1995, and that he then encountered several other families, each with a child who had developed autism or similar conditions after being given the MMR vaccination. This was a three-in-one injection that had first been introduced in the UK in 1988 to replace the original single injections. Along with twelve co-authors, Wakefield published his research in 1998 in *The Lancet* proposing a link between autism, bowel disease and the vaccination based on the cases of twelve patients.

There was a bit of a warning sign here. The sample Wakefield was relying on was alarmingly small. However, Wakefield's treatment of his research was bombastic: he held a press conference calling for the withdrawal of the MMR jab. This was picked up by the tabloid newspapers and fear of the vaccine became widespread over subsequent years, with the take-up rate falling from over 90 per cent to about 75 per cent. There had been over 85,000 cases of measles in 1987, the year before the first MMR injections; then, after a long hiatus, a teenage boy from Manchester became the first person in the country to die of measles for fourteen years, while cases of the disease rose to a twenty-year high. One single outbreak in Wales in 2013 involved over 1,000 cases.

The tide started to turn against Wakefield in 2004 when *The Sunday Times* journalist Brian Deer reported on an investigation into his finances. He claimed that many of the families in the case study had been part of a legal action against the MMR manufacturer and that Wakefield had been paid by the solicitors to give supporting evidence. This led to ten of Wakefield's co-authors withdrawing their support for the part of the paper alleging a link to autism.

In 2007 the General Medical Council started investigating Wakefield and this led to him being struck off as a doctor in 2010: this was for dishonesty, his treatment of developmentally delayed children in giving them unnecessary and invasive medical procedures, and acting without ethical approval for his research. In the same year, *The Lancet* retracted the original paper; the editor Richard Horton said that parts of it were 'utterly false' and said he 'felt deceived'.

Of course, that retraction hasn't stopped the anti-vaxxer movement. Recently it has even moved on to claims that vaccinating your dog can cause 'canine autism' (the existence of which is debatable). The movement continues to flourish, especially in the era of fake news and social media. Wakefield himself moved to America and continued to campaign on the issue, frequently referring to his research while acknowledging it wasn't 'proof' of a link. He continues to point to an increase in diagnoses of autism since 1998, although critics counter that higher awareness of autism disorders has contributed to that rise.

One of the really worrying things about retracted research papers is how often they continue to affect the issue at hand. The website Retraction Watch, which monitors discredited scientific results, lists Wakefield's original paper in its top ten of retracted research papers that continue to be cited in new papers. Indeed, several of their top ten have been cited more frequently after their retraction than they were before.

Bullet Point: As I said earlier, a lie is halfway round the world before the truth has got out of bed.

MAGICAL RESULTS

One of the most prolific scientific fraudsters of recent years was Yoshihiro Sato, a bone researcher in Japan, who fabricated the data for dozens of clinical trials that were reported in journals around the world.

It is something of a mystery what Sato's motivation was and why his activities weren't spotted earlier: he published over 200 papers, which was unusually prolific, and these included many remarkable results.

(As it happens, quite a few scientific frauds have originated in Japan: it might be conjectured that the culture of honour and trust in the country means that scientists aren't subjected to intense scrutiny by their colleagues so the odd bad apple can flourish.)

The fraud was eventually uncovered by dogged work from a New Zealand group of academics: they carefully analysed the studies, checked the statistics and pursued their concerns until it was acknowledged that Sato had indeed been lying about his results. One of the first red flags had been noticed by Alison Avenell, a clinical nutritionist at the University of Aberdeen who ended up working with the New Zealand group, who was reviewing papers to evaluate whether or not vitamin D reduces the risk of bone fractures.

She noticed a weird coincidence in two of Sato's papers. One was about stroke victims, while the other was about Parkinson's disease patients. However, the mean body mass index of both the control group and the study group in each were identical, as though they had simply been cut and pasted. This was a revealing sign, and she noted other similar anomalies on further examination.

It turned out that other scientists had expressed some doubts. In a 2005 paper Sato had claimed that the drug risedronate reduces the risk of hip fractures in women who have had a stroke by 86 per cent, which would be a remarkable result, to the point of being almost magical. Three researchers from Cambridge wrote a patient letter to the journal noting that the study was extraordinary, but wondering how Sato had managed to find 374 patients in just four months. This is a case of a Potemkin number (see p. 2) being the red flag for a liar.

A couple of years later, Jutta Halbekath, a German health journalist, questioned how Sato had managed to recruit 280 patients in two months for a study of male stroke patients, and 500 for a study of women with Alzheimer's disease in a similar period, both of which again showed incredible results for risedronate. On top of this, Sato claimed to have diagnosed all of the cases himself and to have carried out monthly assessments of each over an extended period. Sato's shifty explanation was that the research had actually been carried out at three different hospitals, which for mysterious reasons didn't want to be named. Halbekath noted that Sato had also studied whether or not sunlight, vitamin D, vitamin K, folate and other drugs could reduce the risk of hip fractures, and in every case had produced remarkably positive results.

In 2008 Avenell was working with Mark Bolland, a clinical epidemiologist at the University of Auckland in New Zealand, on a meta-analysis of calcium supplements. Following a discussion of the way extreme results like Sato's could skew a review of various papers, Bolland looked into the papers with his colleagues. He was immediately struck by how improbable the numbers seemed. But he went deeper: when comparing two data sets, scientists can calculate p-values for attributes like age, weight or bone density. This is a statistical measure of the similarity between the groups, and if they are randomly selected it should be anything from zero to 1, with the latter indicating close similarity. But Sato's consistently came in at above 0.8, indicating that he had made up consistently similar sets of figures.

The group had a tough time getting anyone to listen to their concerns. After a two-year wait, one major journal turned down their report on Sato's inconsistencies. A noted journal editor has pointed out that journals aren't fond of these kinds of controversies: it is a lot of trouble and no one is going to thank you for it if you do rigorously go over old work. But eventually, they were able to find a journal that published their work and one by one the studies Sato had published started to be retracted.

As with Andrew Wakefield, the problem is that the initial published claims continue to have an impact long after they have been debunked. Sato's lies continue to be cited in new research; meta-analyses have come to false conclusions based on his research; and huge amounts of effort have been expended by researchers who recruited patients for trials attempting to replicate his results.

Avenell has noted that when it comes to a mere twelve papers that cited Sato's work in major journals, they have been referenced over 1,000 times and over twenty major reviews or meta-analyses have included his work. The use of bisphosphonates to prevent hip fractures in patients with Parkinson's or stroke victims has been encouraged by a single Sato paper reporting eight 'trials'. And Japanese health guidelines issued in 2011 recommending the use of vitamin K for the same purpose were based solely on his work.

Bullet Point: Scientific lies can often be uncovered by patient statistical analysis. But lies aren't easily killed off.

LYING SCIENCE, AN OVERVIEW

The two cases above are particularly egregious cases of scientific deception. And there have been other high-profile cases. For instance, Hwang Woo-suk is a South Korean veterinary researcher who became notorious after falsifying a series of experiments in the field of stem cell research. He was considered a pioneering expert until it became apparent that much of his research data had been faked. And Jan Hendrik Schön is a German physicist who became well known for a series of apparent breakthroughs with semiconductors. As with Sato's research, it was duplicated data that first revealed his deceptions: various other physicists, some of whom had relied on his work, noticed anomalies. One graph had been duplicated in different places (for different purposes). The noise figures for two experiments carried out at very different temperatures were reported as being identical. These duplications suggested problems and, while Schön claimed they were honest mistakes, several of his papers were withdrawn by the journal *Science* and his doctoral degree was revoked.

So there have been some significant scientific deceptions. But how often do scientists actually lie? There have been a few conflicting surveys that have directly asked scientists how often they lie: one meta-analysis* of these surveys found the following.

A weighted average of just under 2 per cent of scientists admitted that they had fabricated, falsified or modified data or results on at least one occasion; 33.7 per cent admitted other questionable research practices. The figures were higher when scientists were asked about whether they thought their colleagues lied: this resulted in an average of 14 per cent for falsification, and up to 72 per cent for other questionable practices. Whether or not this is because scientists were reluctant to admit to their own bad behaviour or because they were paranoid or jealous enough to believe their colleagues were behaving in a worse manner than them must be a matter of conjecture.

There is something of a spectrum when it comes to what constitutes questionable behaviour. The mathematician Charles Babbage once

* https://www.ncbi.nlm.nih.gov/pmc/articles/PMC2685008/

referred to 'cooking' data as 'an art of various forms, the object of which is to give to ordinary observations the appearance and character of those of the highest degree of accuracy'. But what attitude should we take to data mining, when it is used to find a statistical relationship that is then presented as the original target of the study? What about scientists who only publish research when it confirms their original expectations? Or who conceal conflicts of interest?

This kind of grey area would explain why estimates of scientific fraud vary so widely: for instance, we can observe that retracted papers make up about 0.02 per cent of the total, and take that as a benchmark for the minimum level. Or we can note that routine data audits conducted by the US Food and Drug Administration (FDA) during the 1980s found flaws in 10–20 per cent of studies, and 2 per cent of clinical investigators ended up being judged guilty of serious scientific misconduct.

Bullet Point: There are scientists who lie or twist the truth; exactly how many is hard to estimate.

INSECTOGEDDON?

In early 2019, there were headlines around the world claiming that insects could disappear within a century. This was based on the claim that 'insects as a whole will go down the path of extinction in a few decades', which was one of the claims made by Francisco Sánchez-Bayo and Kris Wyckhuys, based on their scientific review of dozens of studies.

Now, it is certainly true that there is some evidence of a decline in insect populations and that this is a real concern. But this was a very strong claim, so it is worth being cautious in seeing how the authors came to this conclusion, which obviously became clickbait due to its apocalyptic implications.

Most experts see the claim as wildly overblown. Elsa Youngsteadt from North Carolina State University has said it's 'not going to happen . . . They're the most diverse group of organisms on the planet. Some of them will make it.' The thing about insects is that there are over a million identified species, and that there are probably millions we don't know about. There are, for instance, more different species of ant than there are of birds as a whole, and more ladybird species than mammal species. And insects are found in an incredible variety of environments, and are quite adept at adapting to changing conditions.

The second problem is that there hasn't actually been that extensive a degree of research into insect populations: there is plenty on certain 'high profile' insects like honeybees, mosquitoes and butterflies, but less on the common or garden creepy-crawly. So there is only a piecemeal set of data on insect populations in the first place.

Sánchez-Bayo and Wyckhuys used seventy-three studies that show insect declines. But they were working from a biased sample, since they apparently found the studies by searching a database using the keywords 'decline' and 'insect'. In addition, the studies were mostly from Europe or North America, while there are huge numbers of insects in other places like Africa and the tropical regions.

So it is hard to draw any wide-ranging conclusions from the studies they reviewed; without good baselines for population sizes, and comparisons from the same locations over time periods, then it is hard to know what kind of pattern we are actually observing.

The review claimed that 41 per cent of insect species are in decline and that global populations are falling by 2.5 per cent per year. But as Michelle Trautwein of the California Academy of Sciences says: 'I understand the desire to put numbers to these things to facilitate the conversation, but I would say all of those are built on mountains of unknown facts.'

So the researchers are not guilty of any kind of scientific misconduct, but they are probably guilty of making inflated claims based on a biased sample. That's not to say that insect numbers aren't declining or that we shouldn't be concerned. But when making apocalyptic claims about the environment, it's best to give a clear indication of how reliable the prediction really is.

Bullet Point: Reviews of studies that have been cherry-picked for 'decline' will probably show 'decline'.

Big Tobacco

When it comes to industries protecting their own business at any cost, and distorting the science in the process, the playbook was really written by the big tobacco companies. During the 1950s, two-thirds of American men and a third of American women were smokers, but there was a slow realisation that there was a link between smoking and cancer. One of the major turning points was the US surgeon general's 1964 report, which referred to 7,000 scientific studies and reports from 150 consultants, and revealed that cancer rates among smokers were 70 per cent higher than for non-smokers. But there had been earlier warnings. In 1953, Alton Ochsner, president of the American Cancer Society and the American College of Surgeons, made a prediction that death rates among the male population in particular would be greatly increased unless a way was found to make smoking less carcinogenic.

As it happens, Claude Teague, an executive at tobacco company R. J. Reynolds, had reviewed the same evidence as Ochsner and reported that 'studies of clinical data tend to confirm the relationship between heavy and prolonged tobacco smoking and incidence of cancer of the lung'. Copies of his report were rounded up and destroyed (step one in lying: suppress the truth). Shortly after Ochsner's statement, six presidents of tobacco companies called a crisis meeting, seeing only a threat to their business. They recruited the advice of John Hill, of the public relations firm Hill & Knowlton. What followed was a strategy described by a lawyer as 'the industry's ultimate public relations sham'.

Hill & Knowlton gave this advice: 'There is only one problem – confidence and how to establish it; public assurance and how to create it – in a perhaps long interim when scientific doubts must remain. And, most important, how to free millions of Americans from the guilty fear that is going to arise deep in their biological depths – regardless of any pooh-poohing logic – every time they light a cigarette.'

They immediately identified a problem, which is that the public was prone to believing scientists and seeing them as neutral commentators. As a result, there was a limit to what could be achieved by advertising, by denying the facts or by directly attacking the scientists who were sending out the message. So Hill suggested a novel strategy: the companies should set themselves up as major supporters of science, calling for more science rather than less.

At the same time, Hill emphasised the importance of scientific scepticism. He suggested the companies should find sceptics of the causal link between tobacco and cancer and amplify their views, and in addition provide funding to increase the number of scientists who were working to disprove the link. By doing this, they could pretend that the controversy over cancer and smoking was an open question, one in which it was equally legitimate to take either side. Only by spreading this kind of doubt could they leave the public in a fog of confusion about the issue, which would in turn allow committed smokers to deceive themselves that the issue was just hysteria. In effect, they had to create 'alternative science' or 'alternative facts' as we might now refer to them.

(I'll return to this subject shortly, but it's pretty obvious now that the fossil fuel industry has followed the same playbook when it comes to climate change denial.)

Parts of the full story were finally revealed decades later in memoirs such as *A Question of Intent* by David Kessler, the former commissioner of the Food and Drug Administration, a long-term critic of the tobacco industry.

Based on Hill & Knowlton's advice, the industry set up the Council for Scientific Research (CTR), which was designed to look like a legitimate scientific body. The director of CTR was a well-known cancer researcher, Clarence Cook Little, who went on to recruit other scientists who were prepared to take the money in return for the work. (Step two in lying: throw up a smokescreen around the truth.)

A hint as to their motivation comes from one of Little's cancer research contemporaries, Charles Huggins of the University of Chicago. He was appalled, sending Little a letter including the lines: 'Please leave the tobacco industry to stew in its own juice ... [it] is criminal to promote smoking. It is dastardly. This is the Age of the Hollow Man. Let it not be known as the age when our finest thinkers sell out.'

The aim was to reinforce the idea that the link was still unproven. They also took advice from lawyers who had noted that in cases against the tobacco industry, they could often influence jurors who saw smoking as a personal choice with the argument that 'association doesn't prove causation'. Of course, this is technically true but still slippery, since the evidence against tobacco was far too overwhelming for it to be a legitimate objection.

The industry also set up the shady body known as Special Projects, through which their lawyers tried to identify scientists and physicians who would testify to Congress against the surgeon general's report. Kessler talked to an anonymous industry source about the lawyers involved in Special Projects.

He asked, 'Where did they cross the line?'

'When you commission the research and know the outcome, that's fraudulent. When you market that as the truth, that's evil.'

(Step three in lying: create a new 'truth'.)

And of course, that was also what was happening when it came to the scientists of the CTR: the studies were carefully set up to produce the desired results, and this continued over the years as the industry used 'low-tar' cigarettes as a way to reassure smokers that they weren't really killing themselves.

As we know now, the misinformation of the tobacco companies continued for decades. But the particular sadness here is the way in which scientists willingly colluded with the industry. This is one of the key moments in which the creation of fake news started to affect the entire national (and international) discourse on a subject, and the continuation of such tactics in other areas by arms manufacturers, big pharma and the fossil fuel industry is continuing to create a world in which alternative facts are used to obscure reality.

Bullet Point: It's always worth knowing where a scientist is getting their funding.

ANNIHILATING THE TRUTH

The phrase 'alternative facts' comes from Kellyanne Conway, who used it to describe lies that the White House press secretary had told on Donald Trump's behest about the size of the crowd at his inauguration. It serves equally well to describe the output of the climate change denial lobby.

Here are some actual facts. The world is warmer than it's been since we started keeping records in 1880. The Australian wildfires of 2019–20 were directly caused by record temperatures and record low rainfall. The Antarctic ice sheet is melting and has contributed to 10 per cent of the global rise in sea levels. The coral bleaching of Australia's Great Barrier Reef, which has been caused by global warming, may now be irreversible. And it is against this background that the Trump administration in America is withdrawing from international treaties aimed at combating the problem and effectively endorsing climate change deniers.

And the climate change lobby is following that same playbook that was established by Big Tobacco. They promote alternative facts, cite a few dissenting scientists in order to create a false sense of equivalency between the two sides of the argument, and throw up a fog of confusion around the subject. The Russian chess player Garry Kasparov has wisely noted that, 'The point of modern propaganda isn't only to misinform or push an agenda. It is to exhaust your critical thinking, to annihilate truth.'

An interesting experiment in 2017 by Sander van der Linden and his colleagues examined how and whether we can combat fake news in general. To start with the researchers gave different groups of subjects different messages about climate change. One group was correctly informed that 97 per cent of climate scientists believe that humans are contributing to global warming. This led them to a reinforced belief in both climate change and scientific consensus on the subject.

Another group was presented with 'alternative facts': they were shown the Global Warming Petition Project, a petition signed by 31,000 people with science degrees arguing that climate change isn't man-made. The problem with this petition is that 99.9 per cent of the signatories aren't experts in climate change, so their opinion is effectively no more useful than the man on the Clapham omnibus. It is a classic smokescreen, using the large number and the word 'science' to present a false impression of statistical significance.

Over a period of six months in 2016, a story about this petition had been the most shared global warming story on social media, which demonstrates how effective such fake news can be.

When the second group was shown this petition, it led to lower levels of faith in climate change and in the idea that there is a scientific consensus about it.

A third group was shown both pieces of information. The interesting thing here is that the result was no overall change in their initial beliefs about either climate change or the consensus: in effect, the facts and the alternative facts cancelled each other out. This is how the truth is annihilated by fake news: given 'two sides of the argument', people don't know what to believe so continue with their existing beliefs, no matter how false they may be.

Possibly the most significant finding of the experiment came from a final group, who were given the facts and the alternative facts. However, they were also given some 'inoculating information': they were informed that none of the 31,000 signatories was a real climate change expert, that while 31,000 seemed a large number it was only 0.3 per cent of the number of US science graduates since 1970, and that the misleading petition had been designed to cast doubt on the scientific consensus.

This neutralised the effect of the alternative facts: the fourth group was almost as reinforced in their belief in both man-made climate change and a scientific consensus as the first group. Given that we have discussed motivated thinking above, it is also worth knowing that this effect was identical regardless of whether the subjects were initially right wing or left wing, and whether they were sceptical of, or convinced by, climate change in the first place. This suggests that this kind of inoculation is crucial when it comes to overcoming motivated thinking.

Bullet Point: In the fake news era, evidence isn't enough; you also need inoculating evidence about the 'alternative facts' in order to protect that evidence.

SOME NOTABLE CLIMATE CHANGE DENIERS

Some climate change deniers are motivated by political beliefs or prejudices. That seems to be the case when it comes to Nigel Lawson, the former Chancellor of the Exchequer under the British government of Margaret Thatcher: he believes that industry should be completely unregulated, so bases much of his critique of climate change (the effects of which he often describes as 'gentle' or 'moderate') on this position. He was the founder of the Global Warming Policy Foundation, which campaigns against climate change mitigation policies and agreements. But it is worth taking a brief look at a few of the other voices in the debate who are more scientifically connected.

Professor Ian Plimer is an Australian geologist. He also happens to be the director of three Australian mining companies. His book *Heaven and Earth* has been described as the 'denier's bible'. Reputable experts debate many of the points he made in the book. One example is his assertion that, when underground volcanoes are taken into account, volcanoes emit more carbon dioxide than humans. This has been debunked by, among others, volcanologist Dr Terrence Gerlach of the US Geological Survey: in fact 130 times more carbon dioxide is produced by man than by volcanoes, including underground ones.

Plimer has also claimed that the evidence for a global flood around 7,400 years ago (which would match it up to Noah's flood) is 'set in stone', a claim that most geologists would regard as overconfident. He is involved with numerous climate sceptic organisations around the world, including the Global Warming Policy Foundation.

Senator James Inhofe was the chair of the Senate Committee on Environment and Public Works in the US until 2017. Between 2000 and 2008, he accepted over $650,000 from oil companies and over $150,000 from coal companies as political donations, making him one of the largest recipients of money from the fossil fuel industry. He has a degree in economics rather than a scientific subject. At one point he issued a list of 400 'prominent scientists' who disputed man-made climate change, but unsurprisingly this was debunked widely as being mainly full of academics who had no expertise in the subject. Inhofe has made many speeches opposing man-made global warming, and has voted consistently for bills that favour the oil industry. During one Senate Committee on Environment and Public Works hearing, Senator Cory Booker mentioned the threat of flooding due to climate change on minorities in his state.

Inhofe contradicted him, referring to testimony from Harry Alford of the National Black Chamber of Commerce (NBCC), saying: 'He provided some of the most powerful testimony that I have ever heard when it comes to the effects of the Clean Power Plan and some of the other regulations ... on black and Hispanic poverty, including job losses and increased energy costs.'

It's interesting to follow this trail through: Alford had justified taking money from the fossil fuel industry on the NBCC's website: 'The legacy of Blacks in this nation has been tied to the miraculous history of fossil fuel ... Fossil fuels have been our economic friend.' At this point, ExxonMobil had donated well over $1 million to the NBCC. In 2015, the NBCC published a report claiming that the Clean Power Plan would 'inflict severe and disproportionate economic burdens on poor families, especially minorities'.

The Union of Concerned Scientists has argued that this report in turn relied on two flawed studies produced by other bodies who had received money from ExxonMobil: the Heritage Foundation, and the US Chamber of Commerce, which received $3 million between 2014 and 2016.

So, it seems fair to question whether the Exxon money provides the missing link here ...

Finally, Dr Benny Peiser is the director of the Global Warming Policy Foundation, and is often quoted by newspapers such as the *Daily Mail* as a global warming expert, in which role he attacks renewables and promotes shale gas and fracking. His academic background was actually as a part-time senior lecturer in sports science at Liverpool John Moores University.

He has also worked for *Energy and Environment*, which has been called the 'journal for climate sceptics' and whose peer-review process has been widely criticised. At one point, Peiser authored a paper critically examining a review by Dr Naomi Oreskes of 928 research papers on climate change. She found they all concurred with the scientific consensus. Peiser suggested that, on the contrary, thirty-four of them rejected or doubted the case for man-made global warming. But later he had to retract this claim and admit that only one of the papers had expressed any actual doubt.

So, dodgy science, weak credentials, dubious numbers and specious arguments ... Would you believe these guys or 97 per cent of climate change scientists? I know which I'd choose.

Bullet Point: Follow the money.

BOGUS CLIMATE ARGUMENTS

Let's take a quick run through some of most common arguments deployed by climate change deniers, focusing on the numbers.

What does the science say?

- 'The Earth's climate is always changing and this is nothing to do with man.' This in itself is clearly true: there have been ice ages and warmer periods throughout history. But it is irrelevant when it comes to the current rate of warming, which is larger than can be accounted for by the natural factors that caused those swings. The effect of carbon dioxide in creating a greenhouse effect has been understood for two centuries and is accepted scientific fact. We know from gases trapped in polar ice that carbon dioxide levels are 35 per cent higher than they have been in the last 600 millennia or so (and, as pointed out above, this isn't 'mostly caused by volcanic activity'). And the rapid increase in global temperatures over the last century is very much what we would predict from such an increase in carbon dioxide. The numbers just don't fit any other explanation.
- 'There are observations of lower temperatures than expected from weather balloons and satellites which don't support the theory of global warming.' There were some measurements in the 1990s that support this objection, but the discrepancies were related to problems with the collection and analysis of the data. For instance, there were variations in how the instruments functioned, and the satellites were sometimes dipping below their calculated orbits. There were also some mathematical errors in the analysis of the data. These have now been resolved

and any remaining discrepancies are within the bounds of error of the predictions made by the models.

- 'Carbon dioxide only makes up a small amount of the atmosphere.' It's only small so it can't be that important, right? Try telling that to someone who just took a small dose of arsenic . . . But seriously, it isn't poisoning the atmosphere, but along with water vapour (which also increases as the climate warms), a modest amount of carbon dioxide in the atmosphere was already helping to keep the atmosphere 30 degrees warmer than it would have been without greenhouse gases. Before the Industrial Revolution carbon dioxide was about 0.03 per cent of the atmosphere, or 280ppm (parts per million). Today, it is about 385ppm. And that increase in the numbers inevitably contributes to further warming no matter how small a proportion of the atmosphere it might be.

- 'The rise in carbon dioxide is caused by increased temperatures, not the other way round.' It's true that increased temperatures can cause a positive feedback loop that means that more greenhouse gases, including carbon dioxide, get into the atmosphere. But by chemical analysis it can be shown that the additional carbon dioxide in the atmosphere is from the burning of fossil fuels, there is no mathematical model that supports the theory it could all have come from positive feedback, and 100 per cent of the data suggests otherwise.

- 'Computer models that predict the future climate are unreliable.' No mathematical models of this sort are perfect, of course. There are simply too many unknowns, and weather is subject to so many micro-factors that it is hard for weather predictions to be made with complete accuracy even weeks in advance. But the models of future climate have been increasingly reliable and coherent over time, and it is hard to find a model that doesn't predict severe future effects of global warming. This kind of argument is specious because it is simply throwing doubt at the science without any alternative model or scientific theory to back it up.

- 'The extent of future climate change and its impact is being exaggerated.' The mid-range future estimates are a global average temperature increase of 2 to 3 degrees centigrade. This would be the biggest change in at least 10,000 years. There will, of course, be varied effects in different regions, but the negative effects of weather events such as heatwaves, bushfires, storms and flooding is certain to have an impact, rising sea levels will cause some areas to become uninhabitable, and the disproportionate effect on poor and developing countries will lead to food and water shortages and, quite possibly, to civil unrest and resources wars. This isn't a moment to be complacent about how bad the numbers could actually be.

There are other specious arguments put forward based on ideas such as 'cosmic rays are causing global warming' or 'it's all about sunspots'. These can be dismissed unless and until someone comes up with a genuine scientific model to explain them. Finally, the most infuriating piece of unmathematical idiocy is on display every time someone claims that we can't have global warming because there's some heavy snow somewhere. Obviously, increasing average temperatures don't mean it will never ever be cold, just that the peak temperatures will be higher, and the lowest temperatures will be less regularly seen in some regions. First prize for this kind of idiocy goes to Donald Trump, who is prone to responding to reports of snow with tweets of this sort:

'Amazing how big this [weather] system is. Wouldn't be bad to have a little of that good old-fashioned Global Warming right now!'

'Brutal and Extended Cold Blast could shatter ALL RECORDS – Whatever happened to Global Warming?'

Bullet Point: The most common arguments against the theory of man-made climate change are unscientific or unmathematical, or both.

HERE'S THE SCIENCE BIT

Finally, to lighten the mood, let's look at a couple of examples of the ways in which bogus scientific facts and numbers are used to sell products. It's probably unfair to blame scientists for this one: it's mainly advertisers who want to exploit the authority bias when it comes to selling beauty products. Think about those anti-ageing creams that are sold with scientific claims like 'This emollient is specially formulated with aqua and humectants', which, ungarbled, simply means that 'this cream contains water and moisturisers'. This is an example of something that merely sounds like a scientific claim, but isn't really saying anything at all.

However, the misuse of science can be more egregious than that. Research published in 2015 in the *Journal of Global Fashion and Marketing*, found that 18 per cent of all statements made in commercials for cosmetics could claim to be trustworthy.

The researchers classified the advertising claims into categories like superiority ('award-winning product'), scientific ('clinically proven'), stand-alone performance ('your skin feels softer'), endorsement from authority ('dermatologists recommend this') and subjective ('all you need for a day of confidence'). They then assessed whether the claim was vague/ambiguous (for instance, 'inspired by science'), a partial truth that omitted information, an outright lie or was acceptable.

They found that just 18 per cent of the 757 claims they looked at were acceptable. Almost half of the claims were scientifically false or subjective claims. So it's best not to put too much trust in the 'science bit' when it comes to beauty products, even if those claims are made by an attractive doctor (or an actor playing a doctor) in a clean white coat.

Bullet Point: Beauty adverts with scientific claims are false, partially false or meaningless 82 per cent of the time.

THE ATKINS DIET

The science of diet is notorious for attracting figures with questionable scientific credentials making dubious claims. For instance, Gillian McKeith is a Scottish television presenter and writer who hosted programmes such as *You Are What You Eat*. While some of her dieting advice is basic common sense, she has supported ideas such as the detox diet and colonic irrigation, as well as claiming she could identify patients' ailments by examining their stools or tongues: all ideas that have little scientific support. She used to style herself as Dr Gillian McKeith but has no accredited qualifications in nutrition or medicine and was eventually asked by the Advertising Standards Authority to stop using the title 'Doctor'.

By contrast, Dr Robert Atkins, inventor of the Atkins Diet, was at least a real doctor. The diet was promoted with the claim that carbohydrate restriction is the key to weight loss. (Other similar diets include the Ketogenic, Paleo, Dukan and Whole 30 diets.) In his books, Atkins asserted that the low-carbohydrate diet creates a metabolic advantage because 'burning fat takes more calories so you expend more calories' and that his diet was thus 'a high calorie way to stay thin forever'.

This is a case where numbers are our friend. A recent study by Sara Seidelmann, a cardiologist and nutrition researcher from Boston, was based on a survey of the eating habits of 447,000 people around the world. Note that the law of large numbers applies here: the sheer size of the study suggests that it is as close to the truth as you can possibly get. It demonstrated conclusively that while banning entire food groups from your diet might have some short-term impact, it has little ongoing impact and can reduce life expectancy and the risk of cancer.

A similar study by the European Society of Cardiology Congress based on 25,000 people in the US (with comparisons to a 400,000+ group of surveys from around the world) was described thus by Maciej Banach, a professor from Lodz in Poland who worked on it: 'Our study suggests that in the long term, [low-carb diets] are linked with an increased risk of death from any cause, and deaths due to cardiovascular disease, cerebrovascular disease, and cancer.'

A third 2018 study based on over 450,000 Europeans over a twenty-two-year period reached a similar conclusion: all the studies showed that the optimal health outcomes are experienced by people who eat modest amounts of meats, dairy, carbohydrates and processed foods while consuming a range of fibre-rich, plant-based foods such as vegetables, nuts, whole grains and beans.

So any extreme diet claiming miracle benefits from complete eradication of a food group is worse for you than simply eating a good old healthy diet. The numbers prove it.

Bullet Point: If you want to lose weight or become more healthy, the only way is to eat less (of a good range of food) and exercise more.

Conclusion: Trust the Numbers

On the whole, science is a noble pursuit, and most scientists are honest and straightforward. The pressures of money, ambition and career security can lead to some scientists crossing to the dark side, and this is a trend that seems to have increased in recent decades. It's notable that when there are lies to be uncovered, it is often patient mathematical analysis or observation that uncovers them. And when junk science such as that of the climate change deniers or the Atkins Diet is on display, the data and the mathematical models are among the best ways of debunking it. So, on the whole, you can probably still trust most scientists. But if in doubt, it's best to put your trust in the numbers.

Bullet Point: The science doesn't lie, but sometimes scientists do.

CHAPTER 7

What's Wrong with Economics?

THE DISMAL SCIENCE

Let me preface this chapter by saying that I am not opposed to economics or to capitalism. I will be saying some critical things about both but I accept from the start that capitalism is, to paraphrase Winston Churchill, 'the worst economic system except for all the others'. What I want to talk about here is some of the inherent problems of capitalism and, in particular, of economics, which is the study of how economies work.

It was the Scottish writer Thomas Carlyle who described economics as the 'dismal science'. (He was provoked either by Thomas Malthus's gloomy predictions of a world of increasing scarcity or by John Stuart Mill's claim that a society's institutions define its wealth.) The problem is that economics isn't really a science at all. I'm going to look into some of the fundamental problems of using mathematics and statistics to study how economies function, but there is a simple underlying problem, which is that economies are a conglomeration of people, each of whom has their own free will, and all of whom have different motivations and behaviours. A science is generally defined as a discipline that studies the natural world and discovers facts through observation, hypotheses and experiments. The fundamental things that a physicist, chemist or biologist studies are unchanging facts: atoms, molecules, cells. But the fundamental building block of economics is people.

So it is hard to carry out economic experiments to test hypotheses. Of course, occasionally there are wholesale experiments such as the twentieth-century attempts to impose Marxism on entire countries (which can be said to have fairly comprehensively demonstrated the flaw in the theory that communism is an ideal system). But in general, economists are studying data that is extremely complex and may be the outcome of any number of inputs, and then trying to summarise this in a few reductionist concepts such as supply and demand.

I've pointed out repeatedly that people are not numbers. And they can't be wholly defined by numbers. So any attempt to reduce numbers to mathematical equations will be in some sense inaccurate or dishonest. But that isn't to say that economics has no value. It's just that it is a pseudoscience, and any time we mistake it for an actual science we are overstating its power to predict the messy, complex elements of human behaviour that are its real building blocks.

The question here isn't so much whether economists lie. It's whether it is fundamentally misleading to pretend that economics is mathematical at all.

Bullet Point: Economics is about human behaviour, not numbers.

A Brief History of Economics

Academics disagree on when capitalism started, but the most common view is that it emerged in northern Europe in the sixteenth to seventeenth centuries. So it is unsurprising that this is the period that also saw the roots of economics as an academic discipline. We'll look below at how the basic ideas of utility, supply and demand were defined and refined in this period. True economics probably started in the early eighteenth century with the French physiocrats, who believed that land – and agricultural land, in particular – fundamentally defined the wealth of a nation.

Their work was one of the inspirations behind Adam Smith's classic work *The Wealth of Nations*, published in 1776. He argued that competition was self-regulating and that governments shouldn't interfere with tariffs, taxes or regulation other than as far as they needed to protect free trade. He laid out some of the fundamental ideas of economics in a clear, appealing way, and remains influential today, especially among those who share his general aversion to government intervention.

The work of both Thomas Malthus and Karl Marx was in some respects a direct response to Smith. Malthus envisioned a world in which we used more and more resources, leading to famine and scarcity. (While he is generally regarded as something of a Cassandra, one could easily make the case that he was just a bit premature in his predictions, given the problems we are now facing with climate change, water shortages, depletion of crucial elements such as precious metals and, especially, phosphorus, whose use as a fertiliser has been fundamental to humanity's ability to produce enough food to sustain itself.) Meanwhile, Marx's entire body of work was rooted in the assertion that capitalism isn't self-correcting in the way that Smith assumed.

The nineteenth-century economists Léon Walras and Alfred Marshall probably did the most to turn economics into a discipline that was explicitly rooted in mathematical models. Marshall gave coherent, persuasive accounts of how to measure and model supply and demand, marginal utility and costs of production, while Walras introduced general equilibrium theory (which is used to show how price is set at the equilibrium of supply and demand) and expanded on the use of statistics in economic modelling.

The two economists that have probably had most influence in the last 100 years are John Maynard Keynes, who accepted Marx's fundamental critique that capitalism doesn't self-correct, as the Great Depression demonstrated, but argued that the role of government in a mixed economy could compensate for this, and Milton Friedman, who returned to a more purist position, similar to Adam Smith, arguing that as GDP grew, the government should consume less capital, instead leaving it in the system so that the economy could flourish without needless interference.

Today we live in a world where there is, in practice, a fundamentally Keynesian approach, with a constant drive towards something that more closely resembles Friedman's desire for less regulation (and, as a corollary, more privatisation). Corporations have, in many ways, as much power as governments, and help to set the political agenda through their lobbying. And economists offer the analyses that help to set political policy; indeed, the art of politics is often carried out through the realisation of fundamentally economic theories.

It is economic theory that is used to define the business cycle, to describe boom, bust and crash, to attempt to control inflation and unemployment, and to influence the behaviour of ordinary citizens. It is the marriage of economics and management theory that gave rise to the whole concept of scientific management, which, as we've seen, was the driving force behind target culture.

So the pseudoscience of economics is one of the key driving forces in the modern world. But how accurate are its descriptions of the human behaviour that underlies every aspect of its field? Let's delve briefly into some of its absolute basics, to see how mathematical they really are.

UTILITY

The concept of utility goes back at least to the great mathematician Daniel Bernoulli's proposed solution to the St Petersburg Paradox, in 1738. This is a thought experiment in which a gambler is offered the following reward for buying a lottery ticket. A £1 coin will be tossed. If the first toss comes up heads, the gambler will receive £2, if the second comes up heads he will get an additional £4, if the third comes up heads he will get an additional £8, and so on, with the additional reward doubling each time. But as soon as the coin comes up tails, the game is over (and their total prize is the money they have won up to that point). The question is how much a gambler should pay for a ticket to this lottery.

The traditional way of answering this question would be to work out the expected value by calculating the average payout. For instance, imagine a game in which you toss a coin and are given £2 if it falls on heads and nothing if it lands on tails; since heads and tails both have a probability of ½, the expected value is (0.5 x 2) + (0.5 x 0) = £1.

In the case of the St Petersburg lottery, the first toss gives a ½ chance of winning £2 so the expected value is £1. The second toss gives a ¼ chance of winning £4 so the expected value is £1. The third toss has a $^1/_8$ chance of receiving £8 so the expected value is £1.

This means that the expected value of the game is £1 + £1 + £1 . . . and as this series continues to infinity, the expected value is also infinite. However, in practice when people are offered a chance to play the game, they are only willing to pay a fairly modest price. So there is a paradox about how to calculate the value of a ticket.

Ten years before Bernoulli's solution, the Swiss mathematician Gabriel Cramer had pointed out that 'the mathematicians estimate money in proportion to its quantity, and men of good sense in proportion to the usage that they may make of it'. And this is the fundamental insight in Bernoulli's solution.

I won't give the full equation as it is complicated, but it relies on the logarithmic function $U(w) = \ln(w)$ (known as 'log utility'), where utility is logarithmically connected to the wealth of the gambler. This is the foundation of the idea that money (and good in general) has diminishing utility. This means that as you get more and more of it, the next unit of money is less valuable to you than the previous ones. If you are a wealthy man, you may think it reasonable to gamble £100 on something that seems like a good bet, but if you are poor and need that £100 for your food and heating, then you will place a much higher value on that same amount of money.

Now, the concept of utility is the fundamental element that underpins supply and demand theory (which we will discuss in the next section). You can draw a curve that estimates how much each new unit of money or each new consumer good is worth to a particular rich person. And you can equally draw a graph that estimates how much it would be worth to a much poorer person. The latter would be much lower than the former. For instance, a rich person might be willing to pay £200 for a cashmere jumper, and only £100 for their tenth cashmere jumper, while someone on minimum income may be unwilling to pay even £20 for one.

To create a demand curve for a group of people, or for a population, you conglomerate all the individual demand curves into one overall demand curve. So it is a concept that is fundamentally rooted in subjective estimates, which are in turn an imperfect, if fairly accurate, way of predicting human behaviour. But this is treated as a fundamental mathematical term that underpins the entirety of the equilibrium theory of pricing.

In practice, utility is defined in terms of how much someone is willing to pay for something: in other words, their subjective valuation of it. It is presumed that this defines how much 'utility' it has for them. There have been some weighty objections to the philosophical basis of this idea. The Cambridge economist Joan Robinson has pointed out the degree to which it is a circular concept: 'Utility is the quality in commodities that makes individuals want to buy them, and the fact that individuals want to buy commodities shows that they have utility.' In addition, she commented on the fact that in order for it to be treated as a mathematical variable, the concept assumes that preferences are fixed, which means that utility is not a testable assumption. We can't really measure utility because we can't be sure whether a change in someone's behaviour is caused purely by the change in price, by a change in their budget constraint or simply by a change in their preferences and beliefs.

None of this is to say that utility isn't a useful concept. But it is worth seeing from the start how vague a term it is, and how it is based on estimates, projections and assumptions.

Bullet Point: Utility is an abstract concept that is often treated as though it is mathematically measurable in terms of money.

CETERIS PARIBUS

When classical economists came up against the fact that it is hard to reduce human behaviour to mathematical variables, they were fond of reverting to the weasel words *'ceteris paribus'*. This is Latin for 'all other things remaining equal'. For instance, let's return to the utility of a cashmere jumper: Jane, who is on a fairly low income, saw a jumper that was £100 last week and declined to buy it, but now she sees it has been reduced to £90, she buys it. Is that because her estimate of the utility of the jumper is that it is worth £90 rather than £100? That would be the assumption made by classical economics. It doesn't take into account other possibilities, such as Jane thinking she is getting a bargain, having seen her favourite TV star wearing the same jumper, or being subject to mood swings or whatever. It's impossible to take all these possibilities into account, but it is OK, because we can simply say that it is all down to her estimate of the utility of the jumper, *ceteris paribus*. In other words, if none of the other things (that we can't measure) have changed, then the change in behaviour is caused by the thing we can measure.

Again, there have been some criticisms made of the *ceteris paribus* concept. The philosopher Hans Albert made the point that the *ceteris paribus* conditions underlying the marginalist theory of demand make the theory into an empty tautology that can't be tested in practice.

Bullet Point: The phrase '*ceteris paribus*' is used to exclude the messy bits of human behaviour in order to turn abstract concepts into something more 'precise'.

SUPPLY AND DEMAND

Again, let's have a bit of history . . .

The basic concept behind supply and demand theory goes back at least 2,000 years. The classic Tamil text *Tirukkural* includes the line, 'if people do not consume a product or service, then there will not be anybody to supply that product or service for the sake of price'. Similarly, the fourteenth-century Syrian writer Ibn Taymiyyah commented that, 'If desire for goods increases while its availability decreases, its price rises. On the other hand, if availability of the good increases and the desire for it decreases, the price comes down.'

In the early days of capitalism, John Locke gave a more detailed account of this idea in *Some Considerations on the Consequences of the Lowering of Interest and the Raising of the Value of Money*. Talking about rent in particular, he wrote that, 'The price of any commodity rises or falls by the proportion of the number of buyers and sellers . . . that which regulates the price . . . [of goods] is nothing else but their quantity in proportion to their rent.'

The terms 'supply' and 'demand' were coined by James Steuart in *An Inquiry into the Principles of Political Economy*, published in 1767. Adam Smith copied the phrase in *The Wealth of Nations* (although he assumed that supply was generally fixed), and other writers who were influenced by him also made it the standard way of talking about this issue. The

French philosopher and mathematician Antoine Augustin Cournot gave what is probably the first mathematical (and diagrammatic) model of supply and demand in 1838 in his *Researches into the Mathematical Principles of Wealth*.

It wasn't until the late nineteenth century that economists properly incorporated the lessons of Bernoulli's solution to the St Petersburg Paradox, as Léon Walras and others adopted a **marginalist** approach, meaning that they assumed that the important factors defining supply and demand operated at the margin (in other words, the factors that would cause supply or demand to rise or fall from their current position).

The first English depiction of supply and demand curves was in the 1870 essay 'On the Graphical Representation of Supply and Demand' by the polymath Fleeming Jenkin, and this was popularised by Alfred Marshall in his 1890 textbook *Principles of Economics*.

Apologies for the detailed history but the notable thing here is how a general, common-sense idea was gradually refined into a pure mathematical concept. We've seen how *'ceteris paribus'* is used in the definition of the demand curve. Similar assumptions are needed in order to make sense of how supply will be affected at the margin.

So supply and demand theory rests on the theoretical quantity of a product that would be offered or bought for a given price, which is an abstraction from real life and one that can't be tested empirically. In other words, it isn't truly a scientific concept.

Bullet Point: The whole concept of supply and demand is an abstraction rather than a scientific measurement.

POSITIVE FEEDBACK LOOPS

There is a really fundamental problem with supply and demand theory. First, here is a basic illustration of the supply and demand curves for ordinary goods. As the price increases, the quantity supplied increases (as producers up production or new producers enter the market), but demand falls as the higher price deters people from buying the good.

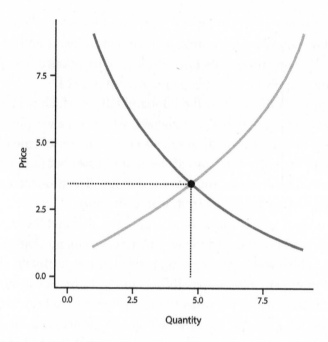

Firstly, let's note that there are a couple of types of goods for which the demand curve is inverse, meaning it slopes upward to the right, rather than down. A Giffen good is one for which, as price rises, consumers still have to buy the good in place of other higher-priced goods they would otherwise have purchased. The classic example comes from Alfred Marshall: low-quality staple foods such as rice or flour. In economies where many people are on a basic staple diet, a rise in the price in these goods will mean that poor consumers can buy less of the pricier goods like meat and vegetables with which to supplement the staple, so will buy more of the staple to maintain their food intake.

The second exception is Veblen goods. These are luxury goods such as champagne or jewellery, for which demand can rise as the price rises. They were first described by American economist Thorstein Veblen in *The Theory of the Leisure Class* (1899), in which he discussed the ideas of status-seeking and conspicuous consumption. In a related category are any goods bought for reasons of snobbery (where the increased price makes the consumer feel more exclusive for owning the good), and goods for which low prices make the consumer suspicious that quality has been compromised, and goods purchased because of a bandwagon effect, where people are becoming more and more aware and attracted to a good at the same time as its price rises. (I've mentioned how the same also seems to apply to anti-ageing creams: see p. 142.)

More importantly, there is the hot-hand fallacy, which is named after gamblers who believe that a sportsperson who is on a winning streak is more likely than they actually are to succeed at the next time of asking. This kind of cognitive error lies at the root of asset bubbles: people see the price of an asset rising, and assume that it will continue to rise, so the price is pushed up further by **irrational exuberance**.

The problem here is that classic supply and demand theory relies on static curves. You can't depict the way an asset's demand changes against price in a bubble with a single curve; instead, this is a situation in which a normal demand curve keeps shifting to the right. For instance, in a property bubble, the houses that would last year have been selling at a certain rate for £200,000 are now selling at a similar rate for £250,000 and people's fear of missing out or greed for profits leads them to value houses higher at every level of the market.

This is a positive feedback loop, which is incompatible with classic supply and demand theory, which works with static curves. Of course, there are still ways of depicting the process, but it is fundamentally in conflict with the assumptions that underpin the theory: that buyers will make rational decisions and that a market that is out of equilibrium will snap back to rational levels of pricing.

To make the matter worse, the simplistic idea that markets are self-correcting has been exacerbated in recent years by the **efficient markets hypothesis**, which suggests that, since everyone has the same information about a market, the market will operate in a rational manner and no one can consistently 'beat the market'. Many of the models that underpin financial analysts' decisions effectively incorporate this assumption.

This has contributed to some of the serious financial crises we have been through. For instance, in 1994 the hedge fund Long-Term Capital Management was founded by the economists Myron Scholes and Robert Merton along with others. Scholes and Merton were two of the minds behind the Black-Scholes-Merton model, a method of pricing options contracts that seemed to make investing in (or gambling on) derivatives considerably safer. It was essentially an equation based on variables such as volatility, type of option, underlying stock price, time, strike price and risk-free rate. They were awarded the Nobel Prize for this work in 1997.

(I'd mention that I agree with the many people who think it a bizarre anomaly that the Nobel Prize treat economics as a science, rather than as a humanity; they don't award prizes for the latter, so by awarding them for economics they are effectively endorsing economists' claims to the status of scientists.)

A year after they won the prize, Long-Term Capital Management got into serious trouble, losing $4.6 billion (£3 billion) in just a few months. The Federal Reserve Bank of New York was forced to organise a bailout of $3.625 billion in order to prevent a much wider meltdown in the financial system. Many complex explanations have been given for what went wrong with their trading systems, including the fact that as they grew they were forced to seek out riskier opportunities. But at the heart of it, there were two problems.

The first was the assumption that the market in general would behave rationally and the other was that the real world would continue to behave like a mathematical model. Any such model relies on mathematical assumptions, whereas the real world throws up outliers or counter-examples on a sufficiently regular basis to make reliance on such models exceedingly dangerous.

Did the financial world learn the lessons of the LTCM collapse? Of course not, they just continued to work with refined versions of the Black-Scholes-Merton model and other similar algorithms. And a decade later the assumptions they were feeding into those models (including the assumption that the entire property market wouldn't fall rapidly in value) exploded, leading to the worst global financial crisis in history.

And how about the regulators: did they learn the lessons of the LTCM collapse? No, they became ever more reliant on their own mathematical models and algorithms, designed by economists and mathematicians to simulate conditions in the real world.

And did the financial world or the regulators learn the lessons of the 2008 global financial crisis?

What do you think?

Bullet Point: Classical supply and demand theory inherently assumes that people are more rational and mathematical than they really are.

Coronavirus Update

The global financial crisis was the worst in history at the time of writing. The aftermath of the coronavirus pandemic will be interesting, firstly as it may cause an even more extreme collapse and also because we will be able to see who profited from the chaos. Recent reports of hedge funds coining it in by betting against the market suggest that the financial sector may not be the biggest loser.

Let's Talk About GDP

As we've seen, the idea of measuring the wealth of nations has been around since the birth of economics. However, the measurement of GDP in particular is a fairly recent invention. In the early 1930s, the economist Simon Kuznets refined it while working for the US Department of Commerce on their strategies for fighting the Great Depression. He regarded it as a technical measure that could not in any way measure well-being or happiness, saying that 'the welfare of a nation can scarcely be inferred from a measure of national income'.

Nonetheless, from the Second World War onwards it became cemented as the main way that national economic activity was measured by the media and the global elites, replacing gross national product (GNP), which had previously been used fairly widely. (GNP includes all income received by companies that are owned by residents or companies of a given nationality, wherever they are based, whereas GDP includes all economic activity in a given geographically defined country, regardless of ownership.)

Its strongest and weakest point is its sheer simplicity. It crams all human economic activity within a country into a single measurement. Strong GDP growth is regarded as a sign of a strong economy and vice versa, and this affects everything from how willing countries are to trade with one another to how cheaply they can borrow money.

So, how is it measured?

There are actually three different methods that, in theory, should arrive at the same figure. You can use the production approach, which adds up the sum of the outputs of every kind of economic enterprise. You can use the expenditure approach, which applies a simple principle: that all of the products must be bought by someone, so the value of the product and expenditure should tally. Or you can use the income approach, which measures the collective income of the 'producers'.

Space doesn't allow for complete detail on how all of these estimates are reached, but here is a quick overview of the elements of the expenditure approach. This involves calculating the cost of the final uses of goods and services, in purchaser's prices, ignoring any intermediate consumption (e.g. purchases made in order to sell on a good or services). The equation for this is $GDP = C + I + G + (X - M)$, which means that we add the sum of consumption (C), investment (I), government spending (G) and net exports (X − M).

Consumption is generally the largest of these components as it includes all household expenditure on final uses of durable goods, non-durable goods, and services. This includes food, rent, household goods, fuel and services such as therapeutic services, but not the purchase of a new house.

Investment includes spending on new houses and business investment in plants, machinery, and so on. It doesn't have the same meaning as in the financial sector, since the purchase of financial products is regarded as saving rather than investment.

Government spending includes the salaries of government employees and all money spent on goods and services that the government will consume or use. Meanwhile, exports and imports are fairly self-explanatory: imports are subtracted to avoid them being counted in the accounts for domestic activity.

The expenditure method is generally reckoned to be the most accurate and easy to measure: clearly all of the elements of it include some degree of aggregation and estimation, but so long as you are using the same methods in different countries and from year to year, it does produce a figure that can be used for comparison.

There is nothing inherently wrong with using GDP to measure economic activity, and it genuinely can capture part of the economic activity in a nation which allows it to spend money on measures that contribute to health, happiness and other public goods. But ideally GDP should be one among a range of indicators. The problem is that it has become the central measure of economic performance, and in this respect it does have some notable faults.

Firstly, a cynic might note how useful GDP can be to politicians of a certain type. If a country gets into increasing levels of debt, and has to spend more money servicing that debt, this would be reflected in GNI (Gross National Income) or GNP but not in GDP. And similarly, if a country sells off its assets and industries to foreign companies or investors, this will also not be reflected in GDP. So if a politician wants to asset-strip a country or to fund their activities with increased national debt, GDP is their friend. GDP doesn't take any account of a country's wealth, only the level of economic activity.

Secondly, since fast growth is regarded as a virtue (and we will look more at this 'virtue' shortly), there are many things that would be regarded as public goods that won't be reflected in GDP and which might thus be regarded as being of lower priority than more measurable economic activity. If you want to focus on issues such as clean air, the environment, equal opportunities or public services, then this will not be directly reflected in GDP whatsoever.

GDP treats all production equally, whether it be the production of life-saving medicine, cars, green-energy devices or nuclear warheads. It is boosted by plastic packaging, which is destined to end up polluting the ocean, and by car crashes (since these are likely to lead to insurance payouts and new purchases). A natural disaster that destroys many buildings is likely to boost GDP. A country's economy is boosted by a building boom that creates a huge amount of new factories, which are never actually used (think, for instance, of the ghost estates that were left behind in Ireland's property bubble, or the zombie factories and empty towns produced in China during the recent economic growth in the country).

It is also blind to quality of life. If the health sector in a country is deteriorating so that all patients are getting worse treatment and outcomes, but expenditure on the health sector has grown, then in GDP terms, that is a 'better' outcome. In the same vein, it ignores the ways that the internet has transformed our experience of life. A free service such as Wikipedia, which makes millions of people's lives easier, is free, so worthless in GDP terms. Peer-to-peer services that allow us to book our own holiday and accommodation are 'bad' in GDP terms since we aren't being forced to pay a middle man for services.

It also only measures transactions that involve an exchange of money. If you buy drugs such as methadone, GDP can pick it up. If you do volunteer work, housework or look after someone in your family who needs caring for, GDP won't notice it. Similarly, it misses large parts of the informal economy, which is a larger part of economies in the developing world, which means it underestimates the true productivity of those countries.

Possibly the most crucial thing about GDP is that it is based on aggregates and (when it comes to GDP per capita) averages. We looked earlier at the use of medians and means when it comes to income: medians are almost always the best measure, but GDP per capita is a mean. So if 0.01 per cent of the population get much, much richer, while 99.99 per cent become poorer, this can result in positive GDP growth. And in general, inequality is not reflected in GDP.

Now, I don't want to tar all economists with the same brush. To be fair, serious economists will rely on a range of measures. Joseph Stiglitz is one notable economist who has been publicly arguing for better alternatives to GDP, while Angel Gurría, secretary-general of the Organisation for Economic Cooperation and Development, has written: 'It is only by having better metrics that truly reflect people's lives and aspirations that we will be able to design and implement "better policies for better lives".'

There have also been some ingenious attempts to find alternative measures that capture some of the important things that GDP misses.

The World Bank has produced some useful studies of countries' actual wealth, which can be viewed alongside GDP measures. In the 1990s, an economist called Mahbub ul Haq, who was working at the United Nations, created the Human Development Index (HDI), which measures life expectancy at birth, adult literacy rate and standard of living and compares this with GDP. There have been numerous suggestions at amendments that make deductions to GDP (and other measures) for pollution, environmental damage and other negative factors. Med Jones, an American economist, introduced the Gross National Happiness Index (also known as the Gross National Well-being Index), which relies on measurements of the quality of environment, education as well as allowing for social and health (mental and physical) indicators. (This was derived from the Kingdom of Bhutan, where gross national happiness is explicitly built into the government's decision-making process.) Stiglitz, along with Amartya Sen and Jean-Paul Fitoussi, has published the results of a commission that investigated ways to adjust GDP to allow for 'well-being economics', which extended the field of measurement to factors such as physical and economic safety and political freedom. The OECD introduced the OECD Better Life Index in 2013. And there are numerous ways of measuring wealth and income inequality (which I will describe shortly). One of my favourite, informal suggestions comes from David Pilling, author of *The Growth Delusion,* who argues that 'average hours of sleep' is a metric that tells us a lot about the happiness, well-being and health of citizens. In the US, the average number of hours of sleep in 1940 was 8. Now it has fallen to 6.8 hours.

So plenty of thought has been put into displacing the central position taken by GDP. But as this point, GDP remains the single most influential metric and the one that has most impact on government policy around the world. Until that changes, and until news networks run regular pieces featuring alternative measurements rather than reporting primarily on GDP, nothing much will change.

Bullet Point: GDP isn't a lie; but over-reliance on it can lead to worrying distortions in governments' behaviour.

THINGS THAT COST BILLIONS OF POUNDS

Think how many times you have heard a newsreader or commentator claim that something has cost the economy billions of pounds. It might be a storm, a major sporting event that leads to people staying home from work, a flu outbreak or an industrial strike.

What is generally meant by this is that the event is predicted to have a negative effect on GDP; in other words, the economy will be less productive. But it is almost always a dubious prediction. Think about a major storm: in the short term, this might mean that businesses lose money. A factory or shop might remain closed, a hotel might remain empty, and people may not make it into the office that day. But in the longer term there are balancing factors. In a town where flood damage has affected houses, there will be costs of repairs, insurance claims and money spent on replacement goods. This isn't good for the householder but it does help to stimulate the economy. In that office people didn't make it into, the chances are that they will have a fixed amount of work to do and will be able to make up the work on following days. People who didn't make it to planned weekend breaks may reschedule, while shoppers who stayed at home are likely to still need the things they might have bought on that particular day.

This is an example of a wider phenomenon: it's always worth paying attention to the way that news defines things as good or bad. When the newsreader announces a growth in GDP, they do it with a smile, whereas the 'this has cost billions of pounds' claim demands a serious face. A rise in house prices gets the upbeat face, while a fall in full-time employment gets the downbeat face. All of these can be seen in different ways, and may be good for some, bad for others: for instance, a rise in house prices is good for those who own property, but bad for those who don't. It might be that such an event is good for the 'average person' but as we've seen there is no such thing.

Bullet Point: Something probably hasn't cost the economy billions of pounds.

Intangibles: Measuring Things That Don't Exist

Even within the measurement of GDP there are some interesting mathematical conundrums: in recent years, the Bureau of Economic Analysis (BEA) in the US saw a need to include 'intangibles' in GDP. This is a term that includes research and development and the creation and development of commercial artworks such as movies and songs, not to mention computer software. The growing role that intangibles play in the global economy has made them a headache for economists around the world, and the US was merely adopting international standards. The problem is how to value research: consider the value a Hollywood studio receives from a movie that is in development for years before being abandoned or the programming Sergey Brin and Larry Page put in when they developed the basic software behind Google. (Incidentally, the latter relies on some interesting mathematics related to the problem of eigenvectors, but that is a story for another time and place.) Clearly one turned out to be hugely valuable and the other turned out to be a money pit. And compare the R&D a pharma company puts into a life-saving cancer drug with the time a rival company puts into releasing a spoiler copy of the same drug.

The best way economists have come up with to measure such intangibles is simply to stick to the expenditure method: assume that the intangibles are worth the money that is spent on them. This is a crude measure, but it helps to iron out the wide variation in actual value, and avoids the need to make any more unreliable estimates. Of course, when it comes to corporate accounting, the measurement of intangibles is a much more slippery creature: it is a notorious opportunity for companies to fudge their books and bolster their assets, by overvaluing imaginary things that might or might not turn out to exist.

Case Study: India's GDP Figures

Obviously, if a country's reputation around the world is partly defined by its GDP figures then there is a powerful incentive for politicians and civil servants to massage the figures. In recent years, the consistently high growth of GDP in countries such as China and India has been questioned by some commentators. For instance, from October 2014 to June 2018, Arvind Subramanian was the Chief Economic Adviser (CEA) to the government of India. In June 2019, he published an article summarising some detailed work he had done on India's GDP statistics, which suggested that they might have been overstated by up to 2.5 percentage points each year since 2011, amounting to a cumulative 19–21 per cent during the whole period (2011–18).

This is hugely politically controversial, and he received much criticism from both sides of the political divide in India. Supporters of the Modi government, who were in charge, were already infuriated by claims that had been published elsewhere doubting the accuracy of their government's statistics, for instance when *The Economist* suggested that India was 'fast catching up' with China 'in the production of dubious statistics'. Government opponents were more perturbed by the fact that Subramanian had only published his thoughts after leaving his post. This isn't the place to establish the truth or falsity of his claims: the more relevant thing is the sheer complexity of that task. So just to give an indication of the debate, here are a few of the arguments that have since been made.

Subramanian's paper was based on certain indicators of GDP growth within the Indian economy and on comparing those indicators across national borders with other countries using statistical regression. The suggestion was that changes made to the way that GDP was calculated in 2011 had introduced systematic distortions.

Now, firstly, other economists have carried out different exercises and come up with different, lower estimates. These relied on metrics such as the ratio between GDP growth and overall credit to the 'private non-financial sector in India from all sources'. The statistician Pronab Sen argued with Subramanian's investigation on the basis that it relied on output variables but didn't allow for productivity gains. The academic Harsh Gupta wrote an article based on the fact that India's tax receipts had been growing rapidly, and that if Subramanian's downgraded estimates were correct, this would make the tax to GDP ratio implausible.

These are just a few examples of ways to debate the issue, and the controversy rages on … The point is that, where GDP statistics are in doubt, it is extremely hard to find a single reliable way to cross-check them, and for any piece of evidence there are counterarguments and possible ways of debunking the evidence. To some degree, total transparency from government statistics departments might help to dispel the fog, but the moral of the story is probably that if any country is indeed manipulating their GDP figures, it will be a pretty hard task to prove it.

Bullet Point: How long is a piece of string?

Coronavirus Update

It is arguable that one contributory factor to the Covid-19 pandemic was news management on the part of the Chinese government. While previous epidemics such as the outbreaks of SARS and MERS had taught us the lesson that transparency and communication between governments is crucial when it comes to controlling a disease, there seems to have been some reluctance (at best) to admit the scale of the problem in China. By the time it became apparent how serious the problem really was, it had already been exported widely around the world and it was probably too late.

The Growth Myth

As I mentioned, the single most significant way in which GDP is used is to measure 'growth', which essentially means a narrowly defined metric of economic growth. Growth has generally been taken as a good thing. But the problem with economic growth is that it is inherently unsustainable. The meltdown in the financial sector that started in 2008 followed a period of huge growth in the industry, which led to it becoming bloated and full of risk. On a wider scale, we have a growing problem of environmental destruction and resource depletion. We might imagine we can keep growing the global economy at an exponential rate, but there is only a fixed amount of the planet and its resources.

Consider what exponential growth means in mathematical terms. If the economy grows at 10 per cent a year, and this leads to a 10 per cent growth in the use of resource X, then in twenty-five years we will need nearly 100 times as much of that resource. There are some resources that are renewable, recyclable or sustainable, but there are plenty that aren't.

The idea that growth is always a good thing is clearly not itself sustainable. It was born in the early days of manufacturing, and of great importance in periods of serious economic instability such as the Great Depression or the post-war austerity years. But, as Joseph Stiglitz among others has pointed out, we are living in a time of different challenges. We are now facing different, existential crises: an inequality crisis, a crisis in democracy and a climate crisis. In the years leading up to the 2008 crisis, economic growth figures around the world were generally strong, and GDP in particular gave us no red flags as to the scale of the problems that were on the way. More worryingly, economic growth since the crisis has been slow but fairly steady and, as Stiglitz says: 'Politicians, looking at these metrics, suggest slight reforms to the economic system and, they promise, all will be well.'

This is all the more reason for us to get beyond the standard metrics of GDP and growth as the accepted approach to assessing a country's economic health. The climate crisis, in particular, is starting to lead to a much wider reassessment of economic health, and a wider acceptance of the idea that sustainability and carbon neutrality will be more crucial over the next century or so.

Bullet Point: The concept of exponential economic growth as an inherently good thing is unsustainable.

MEASURING INEQUALITY

We've seen how GDP is blind to income inequality. There are some useful mathematically based metrics that do measure this. The most widely used is the **Gini index**. Imagine a perfectly equal society in which each citizen is on an equal salary. If you plotted the share of national income received against the number of people, you would get a straight line tilting up from left to right: this is the **line of equality**. Now imagine a society in which one person earns all the income and everyone else gets none. If you started with the poorest member of society and plotted share of national income in the same way, you would get a flat line along the X axis, then a vertical line representing the sudden leap in share when you add the final person, who gets all the income.

Obviously these are the two extremes: most societies lie somewhere in between, meaning that you will usually get a line that curves upward towards the 100 per cent mark. The **Gini coefficient** can be defined as the ratio of the area between the line of equality and the curve, over the area of the entire rectangle. A coefficient of 0 would represent perfect equality, while 1 would represent complete inequality.

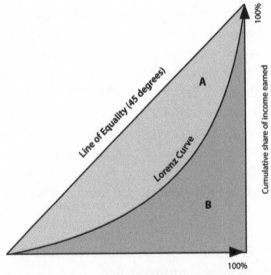

Cumulative share of people from lowest to highest Incomes

In theory, the Gini coefficient can be larger than 1 if we use it, for instance, to measure wealth and count people with debt as having negative wealth. But whether used to measure income or wealth, it will almost always fall in the range from 0 to 1.

There are some problems with the index: one is that it can give the same result for societies with very different wealth distributions. It is also relatively insensitive to changes at the very top and the very bottom of the distribution, which can be the most significant in social terms.

One alternative is the **Palma index**, which is designed to avoid the latter problem. Its origins are in the work of the Chilean economist Gabriel Palma. He showed that middle-class incomes usually account for about half of gross national income while the other half tends to be divided between the richest 10 per cent and the poorest 40 per cent. However, the distribution between the latter two groups varies to a strong degree in different economies. So, to get the Palma ratio we divide the income of the top 10 per cent by the income of the bottom 40 per cent.

In a similar vein, the United Nations Development Programme Human Development Indicators uses the **20:20 ratio** in which we compare how rich the top 20 per cent are with the wealth of the bottom 20 per cent. In recent years, Japan and Sweden have had a low equality gap on this measure, as the richest have been earning about four times as much as the poorest 20 per cent. In the same period the difference in the UK and the US was closer to seven to eight times.

A really simple measure of inequality is the **Hoover index**, also known as the **Robin Hood index**: this measures the proportion of all income that would need to be redistributed to achieve perfect income equality. One way of understanding this visually is to look at the graph above depicting calculation of the Gini coefficient. The Hoover index is equivalent to the longest vertical line that can be drawn between the curve and the line of equality. Like the Gini index, this will range between 0 for a society in perfect equality and 1 for a perfectly unequal society.

Meanwhile, it is also worth considering how income inequality operates within companies and corporations. The **Galt score** is simply the ratio between the pay of a company's CEO and the pay of the median worker. There have been some serious attempts to consider ways of legislating against (or at least shaming) companies in which this index (or the ratio between the highest and lowest-paid employees) is unusually wide.

Bullet Point: Income inequality is relatively easy to measure, but not so easy to address.

When is a Pay Rise a Pay Cut?

There is a bit of a conundrum affecting many people in Western economies at this point in history. Not everyone gets an annual pay rise, but more people do. However, most people find that their standard of living is steadily falling. How come?

The answer is pretty obvious, of course: the rate at which the cost of living is increasing, whether it is housing costs, bills or everyday expenditures, is tending to outstrip the rate at which our incomes are increasing. So it is all down to inflation and the way it filters through into our daily lives.

But what is inflation, and why can't the economists and politicians do something to stop this steady fall in living standards? Let's have a look at the thorny question of inflation and what it really means.

MEASURING INFLATION

The whole concept of inflation is a can of worms. Our general, everyday understanding of what it means is 'the rate at which the cost of living is increasing'.

A more formal definition is that inflation is a way of measuring the rate at which the average price level of a basket of selected goods and services in a particular economy increases over a set period of time, which is usually a year. Mathematically, if it takes more units of currency, whether it be pounds, dollars, yen or whatever, to buy the same amount of goods and services, then each unit of currency is worth that much less.

Economists tend to focus (in a variety of ways) on the impact of money supply on inflation. In essence, money supply is a measure of the amount of money in circulation in an economy. Inevitably, there are different ways of measuring this. The narrowest definition relates to the amount of physical currency in circulation and commercial banks' deposits with the central bank. A more useful definition also includes people's deposits in banks, which includes money that has been created by commercial banks through loans (we'll look at this process more in the next chapter).

You can also use different definitions that include or exclude all kinds of financial instruments that are traded by banks and other financial institutions.

The key thing is that economists tend to agree that sustained inflation occurs when a nation's money supply is growing faster than its economy. Of course, this takes us back to a different can of worms, which is that the latter figure is mostly defined by GDP. But there is an obvious truth in the basic relationship, as can be seen in the economic problems suffered by Spain in the sixteenth century when a sudden influx of gold from the New World made the currency less valuable, and by countries like Zimbabwe in the period when it attempted to counter economic stagnation by printing excessive amounts of currency, which led to hyperinflation.

When it comes to inflation, there is a range of options available: in the UK, the usual measures are the **consumer price index** and the **retail price index.** These are based on slightly different baskets of goods and services, and use different formulae for allowing for the effects of consumers changing their purchasing habits when specific prices increase.

To briefly unpack that, CPI uses the **Jevons formula** to allow for changing consumer behaviour for most goods. The Jevons formula uses the 'geometric mean' of all price changes it is applied to: this essentially means that you multiply all the numbers together then, if there are n numbers, we take the nth root of the product. This will tend to be lower than the arithmetic mean, but it is intended to allow for substitution effects following price rises in particular goods. RPI uses the **Carli method** instead, which does rely on arithmetic mean, and thus tends to be higher than CPI. (The IMF is one of numerous international bodies that advises against the use of the Carli method, on the basis that it produces an upward bias.)

There's no need to get into every detail of the two measures to note one major issue. When it comes to money that the UK government pays out, such as increases in welfare payments, or the salaries of government employees, they have tended to use CPI; when it comes to money that we pay, such as increases in the prices that rail companies are allowed to charge for their tickets, vehicle tax, duties charged on tobacco, alcoholic drinks and air travel, and interest rates on student loans, then RPI has been used. This automatically means that the comparison between those two aspects of our everyday expenditure and income is immediately subject to a gap that is widening at about 1 per cent a year.

In the US, there are two further definitions used: the **wholesale price index** and the **producer price index**. Rather than getting into the details of how they operate, let's take a look at some of the arguments that have been provoked there by the government statistics.

Firstly, there is widespread scepticism in the US about the role of the central bank and the government when it comes to the money supply. Many hark back to events such as Nixon's decoupling of the dollar from gold in the early 1970s, when he said that '. . . your dollar will be worth just as much tomorrow as it is today. The effect of this action, in other words, will be to stabilise the dollar.' The obvious untruth of this statement speaks for itself. But this tends to feed a somewhat inflated fear about inflation and even predictions of hyperinflation. Past legislation such as the Humphrey-Hawkins bill of the 1970s (which rather vaguely called for full employment and zero inflation, without showing much understanding of how that could be achieved) is still regarded as being a promise broken.

It is certainly true that one can pick some alarming statistics on prices of individual goods. New trucks cost about $2,500 fifty years ago, and closer to $50,000 today. Postage stamps have gone from $0.06 to $0.55. The price of cigarettes or gold has increased by 70 to 100 times in a century. All of these instances suggest persistent inflation in the region of 4–5 per cent over those periods. On the other hand, there are plenty of consumer goods such as televisions or computers for which prices have been more stable, or have fallen, which is one reason why it is dangerous to cherry-pick examples to estimate inflation.

Quite a few public figures in the US, including commentators such as Marc Faber and Peter Schiff, are critics of the Federal Reserve's policies on quantitative easing (of which, more in the next section). They suggest that inflation is being understated by the government. According to government figures, inflation has been below the government target of 2 per cent for the last decade, whereas opinion polls regularly show that a fairly significant proportion of the public believe it is over 5 per cent.

I already mentioned one of the prophets of high inflation: John Williams, who runs the Shadowstats website, which publishes his own relatively high estimates of inflation and unemployment. But writers such as John Aziz have pointed to the fact that the methodology used in these calculations is questionable: at one point Williams admitted that he essentially just used inflated estimates of the CPI figures.* This meant that, in trying to prove that the government figures were rigged, he was simply providing a massaged version of the figures that he was casting doubt on and relying on his preconceptions to choose what algorithm to use.

It's worth knowing that MIT, who are independent of the US government, have created their own price index, which is called the Billion Prices Project: it relies on a broad range of data and tends to show that the government data is reasonably accurate. So there are reasons to doubt that the US government is lying as outrageously about inflation as some people believe. But what can be said is that this is another instance in which the problem of motivated numeracy (see p. 6) will mean that a fog of confusion will continue to surround the true rate of inflation in the country.

> **Bullet Point: There are many ways of defining inflation, and some obvious motivations for governments and third parties to choose the method that suits them best.**

QUANTITATIVE EASING

When the global economy crashed in 2008, it was largely the dishonest, overly complex instruments that had been created by the financial sector that were to blame. Essentially, bad, risky debt was bundled up into securities like mortgage-backed securities (MBS) and collateralised debt

* http://econbrowser.com/archives/2008/10/shadowstats_res

obligations (CDO). These were given unwarrantedly high ratings by risk agencies, who were effectively in the pocket of the industry, and thus attracted higher interest rates and better returns than government securities. So to summarise, bad debt was disguised as good debt.

(Scott Adams had a great take on this in *Dilbert*, where one character advises another to invest in diseased livestock: the rationale is that while one sick cow would be an unwise investment, if you take a whole bunch of sick cows 'the risk goes away. It's called math.')

Unsurprisingly, the whole house of cards collapsed. Analysts relying on mathematical models such as Black-Scholes-Merton (see p. 178) hadn't factored in the possibility of a sustained downturn in the housing market. But once house prices had become overinflated, due to lenders feeling secure in offering mortgages to low-income households (since they could parcel up the debt and sell it on), it was increasingly likely that the bubble would collapse. Collapse it did, along with many banks and large sectors of the global economy.

We'll look at one or two of the more egregious examples of bank practices in the next chapter, but for now I want to focus on one of the ways that governments around the world tried to address the resulting crisis.

Quantitative easing (QE) involves central banks increasing the supply of money by buying securities, including government bonds. This is one way of stimulating activity in the economy: following the Great Depression of the 1930s, governments spent borrowed money directly (fiscal policy) in order to create employment and long-term infrastructure. The idea behind QE is that monetary policy can be used rather than fiscal policy: the money banks get from the process can be lent out, so the price of borrowing falls, and, in theory, investment in the economy increases. In pursuit of the same goal, governments also slashed the interest rates set by central banks, in many cases to close to zero.

It's interesting to note that QE had been used over the previous decade by the Bank of Japan, following the Asian Financial Crisis of 1997: the Bank of Japan initially bought government bonds, but moved on to buying private debt and bonds. This was in spite of the fact that the BoJ had previously investigated the issue and concluded that 'QE was not effective'. In the case of Japan, this proved prophetic. GDP continued to fall, from a high of $5.45 trillion in 1995 to $4.52 trillion in 2007 nominal terms.

So why did international governments follow the example of the BoJ? One major reason is that an international consensus had developed that national debt and government spending were bad things. QE doesn't technically increase national debt, since it increases both the assets and the liabilities on the books of the central bank. To some degree this is an illusion: since some of the assets are actually the kinds of instruments that initially caused the crisis, which may never genuinely be saleable.

QE has continued in many economies (for instance, in the UK, the Bank of England announced an additional £70 billion of QE following the Brexit referendum). And low interest rates have also continued to be applied.

Did it work? Up to a point. Asset prices didn't collapse as severely as they might have. Banks were thus able to avoid failure and, by offloading some of their risky securities on to the central bank, they were also able to continue lending, in spite of the new restrictions that were gradually brought in on their asset ratios.

It's arguable that increased government spending actually had more impact on the economy. In the UK, the economy was recovering pretty well in 2009 and 2010, although the austerity policies applied by the Conservative government following a campaign that dishonestly blamed national debt and government spending for the financial crisis (while more or less ignoring the role of the banks) set the economy back considerably. Meanwhile, in the US and other economies, government spending was one of the main stimulating factors that helped in recovery.

The essential deceit of quantitative easing is that it isn't really 'printing money'. You can read on central bank websites around the world elaborate explanations relying on sophistry to explain why it absolutely, definitely isn't printing money. The idea of money printing as a tool for fighting economic recession has been toxic since at least the 1930s, when excessive use of it famously led to hyperinflation in Germany. So governments can't be seen to be doing that. But QE nonetheless is a sneaky way of trying to achieve the stimulating effect of money printing without raising the spectre of hyperinflation.

The stimulation is, however, indirect. From 2008 onwards, the US Federal Reserve effectively increased the money supply by $4 trillion through QE. At the peak of the programme, the banks were still holding $2.7 trillion of that money. So the money had helped banks both directly, by taking dubious assets such as mortgage-backed securities off their hands, and indirectly by making them more stable.

The main effect of QE was to prevent a collapse in asset prices. This is a conundrum: it is clear that asset prices including stocks and property prices were severely inflated by the time of the crisis, but also clear that a severe correction in prices would have led to a far bigger crisis, as more banks would have collapsed.

Governments today explicitly target a steady inflation rate (usually around 2–3 per cent a year). In some respects you would expect the target to be a constant value, but the problem is that deflationary cycles can be far harder to escape than inflationary ones, so there is a logic to the targeting. But when assets become overpriced, there is an argument that it would be best to let the prices fall.

However, a fall in asset prices damages banks who have lent against those assets, and the main owners of banks and assets are, of course, the wealthiest in society. The main beneficiaries of a fall in house prices (for instance) would be the poorest, who struggle to afford the continuing high price of property. So the indirect effect of QE was to allow banks to continue with many of the practices that had led to the crisis, meaning the global economy continues to be reliant on extreme levels of private debt, while simultaneously increasing inequality between the richest and the poorest in society.

The economist Danny Blanchflower has described the thinking behind the populism exemplified by Brexit, les gilets jaunes and MAGA hats thus: 'Those bastards, they got rescued. The banks raised asset prices and the stocks that fell suddenly came back up. If I don't have assets, what's happened is I'm pissed off . . . Brexit and populism is like this: these people aren't hurting, so I'm going to mess with them so they're hurting as well.'

People may not always know exactly what the lie is, but they definitely know when they have been lied to.

Bullet Point: If it walks like a duck and quacks like a duck, it's a duck.

THE ROOT OF ALL EVIL

At its heart, economics is all about money. But what is money?

Historically, money was initially a means of exchange: it's more efficient than barter if you have some kind of token that is widely acceptable. The first types of money were rare or valued goods such as seashells or metal. Over time, coins were used, then notes, which were initially a promise to pay gold or silver on demand.

As notes became more widely accepted, the need to convert them to a precious metal gradually diminished and, since the 1970s, paper money around the world is no longer directly linked to gold. Instead, it is backed only by the authority of central banks, who are in charge of creating it.

Some people distinguish between real money (like gold), which has 'innate value', and fiat or token currency, regarding the latter as inherently unstable and dishonest. The problem with that view is that there is no such thing as innate value. Even a rare metal such as gold has fluctuating value depending on supply and demand and context. If you are stuck on a rowing boat without oars in the ocean and without supplies, you won't value a bucketful of gold as highly as you would a bucketful of water. Gold does tend to maintain a high value, but its value is nonetheless subjective. And when it is used as currency, part of its value comes directly from that use: some people value the metal purely for its transferability (and as a store of wealth) and thus demand for it is increased.

However, there are specific problems with fiat currencies: their supply is much more flexible than the supply of a precious metal, and this is what allows for such phenomena as hyperinflation and quantitative easing.

Anyhow, the essential point is that money is a human construct. Nothing would count as money if it weren't for the way we use it. Money today really is non-existent in some respects. The largest proportion by far is made up of digits in computers. And much of it is created and destroyed by banks when they lend and recover money. (If that doesn't make sense to you, I'll explain it in the next chapter).

The world economy is hugely dependent on money that is loaned into existence, and much of the money governments spend is similarly created by the central bank (or by the creation of government bonds and the like).

For an essentially imaginary or abstract construct, money is hugely powerful. The wealthy have great power, while those without money struggle to afford food, water, healthcare and housing. We treat money as a real thing. But the way it is created and used is largely a business or political decision. It could be created and spent in very different ways if we collectively chose to do so.

You sometimes hear politicians who are averse to public spending deriding their opponents by claiming they believe in a 'magic money tree'. This is a remarkable bit of doublethink, because they know perfectly well that there is such a thing. Whether it be QE or other forms of money creation, money can be wished into existence at the touch of a button. The underlying message of such claims is to criticise or undermine their opponents' intentions when it comes to spending money, whether their priorities are wealth redistribution, social care, defence, policing or whatever.

The lie here is that money can't be created. What these politicians should really say is: 'There is a magic money tree, but I think you should put me in charge of it rather than my opponents.' Because the person who controls the purse strings wields great power and they can use this for whatever good or evil end they want to pursue.

Bullet Point: Money doesn't exist, but it still rules the world.

Coronavirus Update

I wrote the passage above a few months before the pandemic. One thing the Covid-19 outbreak has shown is that governments can find ways to create huge amounts of money when it is an emergency. They didn't just find a money tree . . . they found a whole magic forest. In the UK alone, we have found money to pay businesses to keep employees on, grant mortgage holidays, take most of the homeless people off the streets and into accommodation, support the self-employed who have lost work, write off NHS trust debts, and theoretically to lend billions to businesses who need it. This is still being framed as 'borrowing', although there is talk of 'perpetual bonds', which would be a way of creating that money with very little need to 'pay it back'. (The UK did something similar to pay for the Napoleonic Wars and the bonds continued in circulation until the twenty-first century.) The interesting things once this is all over will be how governments respond: any attempt to return to 'austerity' in the aftermath is likely to conflict with the cold reality of kick-starting stalled economies. And now that it is clear the money can be found to deal with this kind of emergency, will voters expect similar efforts to be applied to other problems: the climate crisis? Unemployment? The social care shortfall? The shortcoming in healthcare systems that were exposed by the pandemic, in spite of heroic input from the workers? Will the world be transformed by the crisis or will it be back to business as usual? My prediction is that attempts will be made to restore the myths that have fuelled the current economic system, but that governments will find surreptitious ways to write off the money they have had to 'borrow'. Whether or not the electorate will be fobbed off with that will be the key thing. By the time you are reading this, the answers may be starting to be apparent . . .

CONCLUSION: WHY ECONOMICS SUCKS

Economics is the pretence that you can reduce human behaviour to a mathematical model. But economics is more than just a description of human behaviour. The mathematical models are then used to justify political choices, and these affect us all. Whether it is the obsession with perpetual economic growth, the measurement of inflation or attitudes to wealth inequality, economics helps to define the terrain on which politicians and financiers operate. So has economics truly allowed us to understand the societies we live in, or helped to create and justify them?

If we live in a world where we are targeting exponential future growth in GDP, then has economics got anything to tell us about sustainability and the future? Of course, there are economists who do attempt to address these issues, but at its core, current economics describes a world run by money, debt and the pursuit of growth. If we want to live in a better world in the future, we might need to look beyond the numbers.

Bullet Point: Money isn't everything.

CHAPTER 8

Financiers and Other Scamsters

....

HOW TO SELL A TURKEY

So, as we know, money makes the world go round. Usury was once regarded as a sin, but now we use QE to stimulate the banks into creating more debt as a way to stimulate GDP growth and inflation. The financial system has gradually become more central to the world we live in.

So it would be nice to think that bankers are all legal, decent, honest and truthful. Unfortunately, there is plenty of evidence to the contrary. In this chapter I'll look at a few of the most egregious examples.

Let's start with Goldman Sachs, the epitome of the 'masters of the universe' of Wall Street. In Karen Ho's book about the anthropology of Wall Street, *Liquidated*, she notes that it is a curious world in which the snobbery networks of Harvard and Princeton interact with an old boys' network, and mathematical geeks and wide boys compete with each other to get tiny advantages in their career. Many have a genuine belief in the 'foundation myths' that they are there to create shareholder value and that Wall Street creates wealth through its wise and careful distribution of capital.

While many of them must thus believe that they are selling a good product, there have been times when it is hard to believe that this was true. In 2007, mortgage-backed securities were all the rage in the financial market. So it wasn't unusual when Goldman Sachs started to sell investors its new Abacus 2007-ACI fund, which was based on the mortgages it had financed. However, it turned out to be a very specific type of fund.

Usually investment banks invest in pools of funds that they expect to be profitable in the future. They also rate the risk of losses and give investors this information when selling the investment.

A credit default swap is essentially a bet on whether a particular investment will rise or fall in value in future. As the asset grows in value, the CDS pays out. With the Abacus fund, Goldman Sachs was selling investors a chance to bet on the future success of mortgage-backed securities. However, Abacus was created for a specific reason. Hedge fund manager John Paulson had analysed the mortgage market and concluded that it was a bubble waiting to burst. Having failed to find a way to bet on it falling, he approached Goldman Sachs. They allowed him to hand-pick a list of mortgage-backed securities that he expected to fall in value. Then they turned around and sold that list to investors as a low-risk investment that would be highly likely to be profitable.

When the market crashed, Paulson made well over a billion dollars, while investors in Abacus saw their money go up in smoke. But don't feel too sad for them: the SEC Enforcement Director did force Goldman Sachs to pay back $550 million when it was revealed that the entire motivation for the creation of the fund was that it would be doomed to fail.

Bullet Point: It takes a lot of chutzpah to not only sell a diseased cow but to also pretend it is healthy.

SHORT-TERMISM

The difference between investment and gambling is rarely defined clearly. My argument would be that, when it comes to stocks and shares, an investor puts their money into a company in order to support it over the medium to long term, but any strategy that involves trying to make significant amounts of money in the short term is gambling.

By this standard, an alarming proportion of modern finance is gambling rather than investment, from flash traders to currency arbitrators and from derivatives traders to the money markets.

We've seen how economists tend to believe they are operating like scientists, gradually building up a body of knowledge about how economies function. This is the main way that economics is taught in universities, using complicated statistical analyses and modelling while failing truly to study human behaviour.

In his 2010 book *The Financial Crisis: Who Is to Blame?*, London School of Economics director Howard Davies pointed out that, 'There is a lack of real-life research on trading floors themselves.' This is just one example of how the mathematical approach to economics has failed to allow for the complexities of real humans rather than abstract 'sellers' or 'buyers'.

The fascinating part comes with the pressures the traders find themselves under. They generally only have themselves to rely on, since they are in direct competition with their colleagues and have to try to keep up with them or beat them. More importantly, they can't rely on their jobs, since there is such a high rate of hiring and firing. In accordance with the most macho strain of scientific management culture, they know that they could be gone at a moment's notice.

Of course, this fosters a fierce desire to achieve short-term results. The inevitable outcome is that traders prize short-term opportunities to make money, rather than developing strategies that lead to more secure, stable, long-term returns.

Which means that the essential conditions they are operating under force them to be gamblers, rather than investors. So their behaviour should be analysed on the same basis as gamblers, and they should be seen as suffering from the same irrational cognitive biases.

The fundamental doublethink here is obvious: while propagating the efficient markets hypothesis, the world of finance is simultaneously turning some of the brightest minds into glorified gamblers.

Bullet Point: If you force people to look for short-term returns, they will inevitably behave like gamblers.

PYRAMIDS AND PONZIS

Two of the most persistent financial scams around are pyramid schemes and Ponzi schemes. To explain the difference between them, here is an example of each.

In 2002 eight women from the Sacramento area were charged with fraud after the collapse of the pyramid scheme Women Helping Women, which they had set up. (I should also make it clear that there is a bona fide women's safeguarding organisation called Women Helping Women:

www.womenhelpingwomen.org.) Women would be invited to parties in beauty salons and homes and promised that they could both help other people and make money themselves. The 'birthday girl' of a particular event would receive up to $40,000 as a gift, which was raised by charging other participants $5,000 each. These other women expected to eventually be the birthday girl themselves.

The fundamental instability in a pyramid scheme is that it is mathematically impossible to keep on recruiting new people. If it takes the recruitment of eight people to fund one women, then for the next round of birthday girls, the number of participants needed will be $8^2 = 64$. And for each of those sixty-four women to become the birthday girl it will need $8^3 = 512$. At each stage, the number grows exponentially larger and eventually, no matter how big the target market, you run out of potential new recruits, at which point the scheme collapses. It's worth noting that people sometimes create pyramid schemes without understanding this mathematical inevitability. The Women Helping Women concept has had a surprising survival rate as it keeps being revisited by women, some of whom believe they really are taking part in an empowering movement.

Like pyramid schemes, Ponzi schemes favour the early participants, but they work in a slightly different way: essentially, the investments of later investors are used to pay out profits to earlier investors, and again this can continue to work until the people running the scheme run out of new investors. They are named after Charles Ponzi: a Boston-based businessman, he issued notes that were payable in ninety days, promising to pay 50 per cent interest. He used the cash of new investors to make the payouts, and made $15 million from his commission on the notes, before he was arrested and jailed in 1934.

In India, the so-called Great Plantation Scam of the 1990s was a Ponzi scheme. Anubhav Plantations was a Chennai-based company founded in 1992 selling shares in teak plantations, promising guaranteed profits both in annual interest and through an eventual payout after twenty years when your teak was mature. The company diversified into four separate companies, all operating in a similar way. However, the company had hugely overstated its actual holdings of plantations and the costs of raising the teak and was paying out new investors from the early capital raised. It collapsed suddenly in 1998, with major repercussions for many investors.

The largest Ponzi scheme in history was founded by Bernard Madoff, who would become non-executive chairman of the NASDAQ stock market before he was exposed. The fraud was estimated at having been worth about $65 billion.

Madoff founded a penny stock brokerage in 1960 and this eventually became Bernard L. Madoff Investment Securities, a hugely successful business that supposedly didn't use 'specialist' firms to make orders, but directly executed them over the counter from retail brokers. An advantage for Madoff of this modus operandi was that it reduced the transparency of the business and limited how many people knew what trades he had actually executed.

For years, he employed his brother, niece and two sons at the firm, which paid out consistently high returns over a long period. These were in the region of 1 per cent to 2 per cent returns per month: the fund was in positive territory 96 per cent of the time, and it produced a 45-degree curve of profit with minimal volatility.

Did anyone notice? Well, Harry Markopolos, a small-town fund manager and maths geek, did. He spent years trying to raise the alarm, having analysed Madoff's results and realised they were implausibly good. He warned the Securities and Exchange Commission (SEC) as early as 2001, and went so far as presenting them with a detailed dossier in 2005 called 'The World's Largest Hedge Fund is a Fraud'.

But Madoff was a respected operator on Wall Street and the SEC wrote Markopolos off as a crank. After the fraud was finally revealed, Markopolos revealed that he'd talked to plenty of other people who had suspicions. Some had believed Madoff was merely using insider trading, but there were some obvious signs that something was amiss, including the fact that he had his books audited by an obscure two-man accounting firm, Friehling & Horowitz.

Markopolos has said that, 'Hundreds, if not a few thousand people knew. They knew something was wrong with Madoff and they stayed away from him ... You'd hear people say, "I don't think he's legitimate, I think he's a fraud."' They were right. In 2008 Madoff's sons reported their father to the authorities: he had finally confessed to them that the asset management unit of his firm was nothing but a Ponzi scheme, and that it was 'one big lie'. He later said that the fraud had started in the mid-1990s, but investigators suspected it had been going since the 1980s. His investors lost all their money, and he was sentenced to 150 years in prison for the fraud.

Bullet Point: If you're offered a sure-fire investment opportunity, do the maths for yourself.

DOUBLE IRISH WITH A DUTCH SANDWICH

Let's not forget the accountants. Earlier in the book I mentioned the role of Arthur Andersen in signing off the accounts of Enron: in that case they were effectively endorsing a dishonest business practice. But there are legitimate ways of cheating the system, especially when it comes to tax avoidance. The double Irish with a Dutch sandwich was a tax avoidance technique that was developed by the accountancy departments at some of the largest corporations in the world. It used Irish and Dutch subsidiary companies to shift profits from American companies to jurisdictions where tax rates were much lower.

This (along with other similar schemes) is a common strategy for tech companies, because it is relatively easy to transfer intellectual property rights to companies abroad. In the double Irish with a Dutch sandwich, profits are sent through one Irish company, on to a Dutch one, and then on to a second Irish company, which has its headquarters in a tax haven. The first company receives large royalties from US sales, which significantly decreases the taxable profit there. The second Irish company performs a similar role, when it comes to sales to European customers, before transferring the profits to the first company via a Dutch intermediary.

It has been reported that, in 2017, Google transferred around $20 billion this way, with the money ending up in the hands of an Irish company in Bermuda.

The scheme was investigated from 2014 in the EU and the Irish eventually legislated against it following a promise made in the 2015 budget. However, companies currently operating the scheme were allowed to continue to 2020, meaning that at the time of writing it is still in operation. And while this loophole has been closed, you can be sure that many of the other sophisticated ways of hiding profits and magically discovering them in a lower tax jurisdiction will continue to be exploited in the future by some of our most loved corporations.

Bullet Point: If you have a load of money to hide, employ a weasel in your accountancy team.

CORPORATE WELFARE

Privatisation is often justified on the basis that the private sector is more efficient than the public sector. One thing that is not mentioned so often is how much the private sector is directly supported by the government. Corporations lobby governments for favourable tax rules, business environments and large contracts, and it is often the largest corporations who benefit the most. For instance, in the United States, there are extensive agricultural subsidies that are justified on the basis that they help farmers to keep their businesses afloat. However, by far the largest part of the subsidies go to large agribusiness corporations such as Archer Daniels Midland or the richest segment of the farming community. One irony of the Brexit debate was that Dominic Cummings, a leading light in the Vote Leave campaign who went on to become Boris Johnson's closest adviser in government, was the co-owner of a farm that had received €250,000 in farming subsidies from the European Union he was so determined to leave.

One analysis of UK government spending found that subsidies, direct grants and tax breaks given to big business add up to £3,500 a year from each UK household, £93 billion in total.* In the US, two professors at the

* https://www.theguardian.com/politics/2015/jul/07/corporate-welfare-a-93bn-handshake

University of Iowa, Alan Peters and Peter Fisher, calculated that state and local governments provide \$40–50 billion annually in 'economic development incentives'.

The government money isn't restricted to companies from the same country; in 2012 Amazon were widely criticised for paying no tax in the UK due to their complex structure, but in the same year they were given £16.5 million by the governments in Wales and Scotland to help them build warehouses. In Wales, an additional £3 million was spent building a road to link the warehouse to the road system.

Money spent by the UK government includes grants to train operators and defence firms, amounts allotted to businesses to allow them to write off billions of pounds spent on plant and equipment, exemptions to tax for the construction sector, and over £8 billion of tax breaks on fuel for airlines. Some of the spending has obvious purposes, such as the money spent helping energy companies to decommission nuclear plants and the state insurance scheme that gives an export credit guarantee to business. But often the reasoning is harder to follow, especially when it comes to the uncompetitive granting of major contracts to private companies for providing services the state could provide, such as the probation system: eight private firms were recently paid royally to run twenty-one 'community rehabilitation companies', but the outcome was so disastrous the whole thing has had to be renationalised, with the contracts being terminated two years early.

Now, having talked about the banking crisis of 2008, what should we think about the bailouts that saved the banking sector? In the UK these were estimated to have cost about £35 billion by 2013. In the US, over \$400 billion was committed to buying distressed assets from banks, and while the Troubled Asset Relief Program (TARP) initiated by Secretary of the Treasury Henry Paulson eventually earned a \$15.3 billion profit, it constituted a significant loss when adjusted for inflation. Zero-interest loans from central banks around the world to shore banks up can also be seen as a form of corporate welfare.

If you or I get into trouble by taking on too much debt and being unable to earn the income to repay the loans, we will eventually have no choice but to go bankrupt. Happily for the banks, they are seen as being 'too big to fail' so they had the option of going cap in hand to the governments to ask to be bailed out.

One of the lies that has been spread about this is the insistence by banks that didn't receive the bailout that they didn't benefit from government money. Since the banks were all so interdependent in terms of loans and debt, the collapse of the banks that did get bailed out would inevitably have led to further failures elsewhere. So there is no sector of the banking system that didn't benefit from this corporate welfare.

As someone wise once said: If you owe the bank £100 it's your problem. If you owe them £1 billion then it's their problem.

Bullet Point: We live in an age of socialism for the rich, free enterprise for the poor.

INEQUALITY ISN'T INEVITABLE

You will often hear the argument that it is inevitable that wealth inequality will continue to rise. The theory is that it is impossible to address this through redistributive taxation because the richest individuals will leave the country or find ways to conceal their wealth in tax havens. Additionally we are told that high taxation discourages 'wealth creators', so would lower living standards all around.

For the moment I'll ignore the fact that the term 'wealth creators' is often applied to individuals such as bankers who arguably should be seen as debt creators and parasites on wealth creation. Instead, let's look at how the orthodox opinion about this subject has changed over the decades.

In the post-war period, up to about 1980, the mainstream political belief was that it was possible to combat rising inequality through taxation, and the policies that were adopted in many countries did indeed lead to a fall in inequality from the 1940s to the 1970s. Since 1980, in both the US and the UK, the share of national income earned by the top 1 per cent of earners has more than doubled, while the average earnings of the bottom 90 per cent adjusted for inflation have barely risen. The earnings of a CEO have risen on average from twenty times the average employee of the firm to 350 times as much.

However, this isn't a uniform global trend. In Canada and Japan, the increase in inequality has been much smaller, while in many European countries, including France, inequality has been stable or has even fallen further. So the rise in inequality in economies such as those of the UK and US can't be an inexorable outcome of global forces.

Warren Buffett, the Sage of Omaha, has described the situation thus: 'There's been class warfare going on for the last twenty years and my class has won.' Higher tax rates have fallen in many countries since 1980, but the changes are most extreme in the UK and US. During the Thatcher government (1979–90), the top rate of British income tax fell from 83 per cent to 40 per cent. Under President Reagan the top US rate fell from 70 per cent to 28 per cent. In both countries, it has crept up slightly since then, but nowhere near the standard post-war rates. This reflects a wide-spread orthodoxy that high tax creates a disincentive to wealth creation and hard work.

The fundamental theory often used to justify this is the **Laffer curve**. This originated in a 1974 meal attended by Donald Rumsfeld, Dick Cheney and the right-wing economist Arthur Laffer. President Ford had recently announced a rise in tax rates; while discussing it, Laffer commented that a 0 per cent income tax rate would raise no income, but so would a 100 per cent rate, since no one would have an incentive to work. To illustrate this, he apparently doodled a curve in a graph on the back of a napkin.

There is some basic common sense in this idea: at the extremes, clearly the 0 per cent rate would raise nothing, and at 100 per cent rates, then whether or not people worked would be purely down to how much they wanted to. So the idea that there is a curve of some sort with a point on it where the tax take would be maximised also makes some sense. But this was an off-the-cuff thought and there has been very little research that substantiates it or convincingly establishes the point at which the tax take would fall. Even economists specifically employed by the Reagan administration to justify the tax cutting policy failed to come up with any evidence for the theory.

However, the Laffer curve is frequently invoked as though it is a well-established economic theory, and used to justify tax cuts at any level. For instance, after the UK top rate had been raised to 50 per cent under New Labour, the Conservative chancellor George Osborne claimed that his cutting back to 45 per cent would increase rather than decrease the tax take (there's no evidence that it did). He was relying on a fairly shaky analysis that suggested that the maximal tax take in the UK would occur at a top rate of 40 per cent, but economists such as Thomas Piketty have suggested that the sweet spot might be closer to 85 per cent. This is partly because of some of the common-sense issues that Laffer ignored, including the fact that most people aren't in jobs where they have an easy choice of simply working harder or longer, while some people will even choose to work less if they can earn the same amount by doing so.

It's interesting to note how much our perceptions of inequality are affected by our idea of whether wealth is usually earned through talent and hard graft as opposed to luck. In countries where a higher proportion of the population believe that luck plays a large part, there is more support for higher taxation and vice versa.

For instance, in the US there is a widespread belief that many rich people have become rich by pure hard work and that anyone could emulate their success, so the poor are poor because they are lazy. This faith in social mobility is not reflected in the facts: the poorest 20 per cent in the US work as hard as the poorest 20 per cent in Europe, and there is no marked disparity in how hard the rich and poor sectors of society work. And social mobility is lower in the US than in many European countries.

We've previously discussed motivated numeracy; there is also a lot of psychological research into motivated political beliefs. It is arguable that high levels of inequality actually lead to the belief that the rich are harder working rather than luckier than the poor. In the US, post-tax income inequality is high and with welfare payments fairly limited there is a pressing need to believe that hard work will rescue you from poverty and a great motivation to try to do that. A similar ethic applies to some degree in the UK, but Europe in general has a more generous approach to welfare, based on the idea that poverty is not always the fault of the poor.

Of course, there is a mirror image of this belief in the super-rich, who often believe themselves to be fully deserving of their wealth and rule out any role that luck may have played. When you see newspaper reports claiming that hundreds of Britain's wealthiest entrepreneurs would 'flee the country' if a Labour government were to raise tax rates, the likes of Andrew Lloyd-Webber complaining about the possibility of a 'Somali pirate-style raid on the few wealth creators who still dare to navigate Britain's gale-force waters', the magician Paul Daniels threatening to leave the country if Labour won the 1997 election (they did and he didn't), or Stephen Schwarzman, CEO of US company Blackstone, comparing tax rises to the Nazi invasion of Poland, you are witnessing a supreme sense of entitlement and self-justification.

As ever, it is worth listening to Warren Buffett on whether or not high taxes would lead to rich individuals fleeing the country: 'Imagine there are two identical twins in the womb ... And the genie says to them: "One of you is going to be born in the United States, and one of you is going to be born in Bangladesh. And if you wind up in Bangladesh, you will pay no taxes. What percentage of your income would you bid to be born in the United States?" ... The people who say: "I did it all myself" ... believe me, they'd bid more to be in the United States than in Bangladesh.'

> **Bullet Point: It's a bad idea to structure your economic orthodoxy around an idea scribbled on a napkin, and a good idea to listen to Warren Buffett.**

SPOOFING THE ALGOS

Algo (short for algorithmic) trading is the use of computer programs to trade stocks and financial assets based on predefined criteria. In theory this can help to stabilise markets as the algorithms takes advantage of small price differences and ensure the market is as competitive and liquid as possible. High-frequency trading (HFT) is the Wild West of algo trading, in which orders are placed by the algorithms at extremely high speeds, taking advantage of even tinier discrepancies in price levels.

Obviously, HFT is much closer to gambling than the traditional model of long-term investment in stocks. One of the dangers this poses is the systemic risk of algorithms creating downward or upward spirals, which is amplified across markets since the algos are constantly comparing pricing in different markets to look for tiny opportunities to exploit for financial gain.

There are numerous problems with HFT, not the least of which is the temptation for their users to manipulate the market via false information in order to create the kind of short-term volatility that allows them to make profits. Many traders use stop-loss orders to protect themselves from sudden downturns in the market: for instance, the order might be 5 per cent below current trading prices. If the market falls by more than 5 per cent those orders are triggered and the stocks are sold at a 5 per cent loss. But when market volatility is heightened, these orders can be triggered by a sudden lurch down in a stock, which is corrected shortly afterwards.

There is also considerable danger in using algorithms that operate at high speed if an error is input into the system. The best known example is the case of Knight Capital, who lost $440 million (which subsequently led to them being bought out) in forty-five minutes on 1 August 2012. The algorithm made millions of errant trades across a range of stocks, buying them at the higher price and selling them instantly at the lower price. The analysts couldn't isolate the problem fast enough and the 'Knightmare' effectively finished the firm off.

There was an even more severe incident on 6 May 2010: the Flash Crash in which the Dow Jones dropped by nearly 6 per cent (or 1,000 points) in a matter of minutes before rebounding. A range of stocks fell by up to 15 per cent while 300 securities were traded at prices that were 60 per cent below their initial value (these trades were later cancelled but many investors who lost money through stop-loss orders in the 5–15 per cent range of losses were not refunded).

The SEC and the Commodity Futures Trading Commission report into the Flash Crash initially blamed a single $4.1 billion trade by a Kansas mutual fund trader, but subsequently the authorities charged Navinder Singh Sarao, a London day-trader, with market manipulation.

The claim was that he had been 'spoofing' the market. This means placing a huge number of orders in an asset or derivative but then cancelling the trades before they are executed: this influences the price in the short term as it gives other traders (including the HFT algos) a false impression of a sudden burst of selling or buying interest. If the spoofer has created fake sell orders, the price will plunge, then they can cancel the orders and instead place a large number of buy orders, pushing the stock back up, before quickly selling at something close to the original price.

The case of Navinder Singh Sarao was an interesting one. His lawyers argued that he suffered from an extreme form of Asperger's syndrome (now classified as a type of autism) and was treating the trading as a mere computer game, one for which he had found a remarkable 'cheat'. He was under threat of extradition and jail time for five years, but eventually he was given a brief non-custodial sentence in recognition of the particular circumstances.

Not everyone is convinced he was solely responsible for the Flash Crash anyhow; while further measures have been put in place to have 'kill switches' that mitigate the risk of such incidents, the fear remains that someone with a more sinister motivation than Sarao could either use HFT algos or exploit their flaws to create significant economic damage. Meanwhile, there are probably subtler examples of strategies such as spoofing being used on a regular basis. Sarao's mistake was possibly to be too obvious in his spoofing. The really smart liars are the ones who are still out there.

Bullet Point: Machines are only as gullible as the people who programmed them.

CHARTISM

If you are looking for investment advice, there are plenty of mathematically dubious ideas to be found. There are two main types of investor. First there are those who believe that markets are irrational enough that it will be rewarding to try to find ways to analyse stocks to find under-priced ones on the assumption that eventually catch up. The second type assumes the market is relatively rational so the 'trend is your friend': all the information you need to know is there in the price and the pattern of the price movements. The latter belief sometimes fosters the essentially irrational method of 'Chartism'.

This involves trying to identify patterns in the charts on the basis that certain types of shape indicate future movements. For instance a 'candle-stick' (made up of the daily high and low, with a closing price marked in) can be used to predict tomorrow's movements. 'Moving averages' are calculated by working out the average closing price over a period of time, then adding the next day or week into the calculation: a rising moving average is seen as a bullish sign, while the opposite is a bearish sign. Channels and resistance points are discovered by drawing 'tramlines' joining up the peaks and troughs of a stock price over a period of time, to look for upward or downward momentum (and future points where the tramlines may meet). And a 'Fibonacci' retracement is supposed to indicate the recurrence in stock prices of the Fibonacci patterns that often appear in nature (in, for instance, the pattern of a flower's seeds and leaves).

One could spend a lot of words trying to explain how fundamentally unmathematical this all is, but there is a far simpler point that can be made. In Chapter 4, we looked at all the different ways that trend lines can be drawn on charts, and how subjective a process it is. I've also mentioned that people are really bad at understanding randomness so tend to impose non-existent patterns on random data. Even investors who are skilled in regression analysis will be hampered by the different interpretations that could be taken from carrying out the analysis for arbitrarily chosen periods of time. So anyone relying on such methods is overwhelmingly likely to identify patterns that aren't really there, and to be misled into making dubious predictions.

Think about a simple comparison. If you took the average growth in women's football audiences over recent years and compared it to the average growth in men's football audiences, you would almost certainly project a chart that shows the audiences for women's football overtaking men's audiences in a decade or so.

But we can easily see the nonsense at work here: women's football is at an earlier stage in its popularity and thus is growing at a faster rate. This will almost certainly level off at a certain point, so the difference in the two measurements gives us little to no information about future growth.

When it comes to stocks there will often be factors we can't so easily identify that lead to short- or medium-term volatility and movements, which won't be replicated tomorrow, next week or next year. So looking for patterns in the data is a fool's errand if you don't also have a significant amount of knowledge about the relevant company and its competitors.

Bullet Point: There's no magical maths shortcut to choosing investments wisely.

WHEN THE LIES ARE UNCOVERED

Having accepted that there are probably still liars out there, let's briefly recap a couple of cases in which bankers were actually convicted for their market manipulations. Nick Leeson came from a fairly modest background and, having failed his Maths A level,* he rose fairly rapidly through the ranks of Coutts and Morgan Stanley before taking up a management position at Barings Bank, the UK's oldest merchant bank. Ironically, one of his early tasks was to investigate a case of fraud in which a bank employee had used a client's account improperly until margin calls from the clearing houses exposed his scam.

Leeson had himself committed fraud, having failed to report a legal judgment against him on his application for a broker's licence and this wasn't disclosed by either him or Barings when he was appointed manager of their new futures and options office in Singapore in 1992.

From the start he made large, unauthorised trades: at first the gambling paid off as he made £10 million profits for the bank (earning a bonus of three times his salary in the process). But his luck turned and he took to using one of the bank's 'error accounts' (which are supposed to be used to correct mistakes made in trading) to hide the losses. Barings made the mistake of allowing him to be chief trader while settling his

---

* To be fair, he did pass his other two A levels, so this is a bit of a low blow from me . . .

own trades, which would usually be separate roles. He claims he didn't use the gambles for his own enrichment, but it was reported in the press at one point that approximately $35 million had been identified in accounts tied to him.

Leeson made two classic gambler's mistakes. Firstly, he threw good money after bad when his luck turned. And secondly, he used a version of the **Martingale strategy**. This involves doubling your gamble after a losing bet. It originated in roulette, where a bet on red or black pays out nearly 50 per cent of the time. The theory behind the Martingale strategy is that if you lose one chip on round one, you bet two on the next round. Then if you lose, you bet four on round three. In theory you will always end up with a profit of one chip. However, the exponential nature of the scheme means that the amount you are gambling escalates rapidly if you have a losing run. After six losing bets you need sixty-four chips. After twelve losing bets you need 4,096, and so on.

Leeson was hiding £2 million in losses at the end of 1992, and this ballooned to £23 million a year later, then £208 million by the end of 1994. All the while he was able to maintain his reputation as a trading genius by reporting successful trades in the books, while hiding the disasters in the error account. Eventually, something had to give. In a desperate attempt to turn the tide, he placed a large bet on 16 January 1995 that there wouldn't be a significant movement in the Japanese stock market overnight. On the morning of 17 January there was a major earthquake in Kobe, which led to a big fall in the markets. After losing even more money betting on a quick recovery in the markets, Leeson left a note on his desk saying 'I'm sorry' and fled the country. The eventual losses of £827 million led to Barings becoming insolvent.

Leeson served a sentence for his role in the bank's collapse, but it has been widely observed that the bank itself was at fault, as its auditing processes were so wildly inadequate and the excessive trust it placed in its employees played a major role in allowing Leeson to continue his misdemeanours for such a long time.

So banks around the world learned to tighten up their procedures and to do everything by the book, right?

Well, of course not, as the Libor scandal demonstrated.

Libor is an average interest rate that is used a benchmark: it is calculated based on submissions from major banks around the world. It became apparent after the global financial crisis that banks had been falsely inflating or deflating their rates either to make a profit or to make themselves appear more secure than they really were. The Libor rate directly affects the gigantic derivatives market, which is worth over $300 trillion, so this was a hugely significant problem.

At the time of writing, numerous criminal settlements by Barclays Bank demonstrated that there had been a significant amount of fraud and collusion among the banks when it came to their Libor submissions, and Bob Diamond, chief executive of Barclays, resigned in 2012 after the bank was forced to pay a record fine for influencing the rate improperly. But only one person has been jailed for their role in the scandal.

Tom Hayes was a successful trader for UBS and Citigroup in Tokyo. However, when Lehman Brothers got into trouble in September 2008, Hayes' trading book was also in trouble. He had huge amounts of money riding on his bet that interest rates and the yen would remain stable, but the situation at Lehman suggested that the rates would shortly be spiking sharply. His only hope was to get out of all his positions before that happened.

When Hayes had been a trader in London, he had got to know several of the people who made the daily submissions for Libor. He had realised how dependent those decision-makers were on interdealer currency brokers. So he sent messages to several brokers he had managed to befriend. To one he wrote: 'Cash, mate, really need it lower. What's the score?' He continued to do this over a three-day period in which his book went from $20 million in debit to $8 million in profit. He still had work to do, though, so he messaged a contact at ICAP, the world's biggest interdealer broker, which issued a 'Libor prediction' email every day, asking them to get their predictions lower: every tick (which is a hundredth of a basis point) in the yen Libor rate made about $750,000 difference to his profits. He sent another broker in London the message: 'I need you to keep it as low as possible, all right? I'll pay you, you know, $50,000, $100,000, whatever. Whatever you want, all right?'

Hayes was eventually given a fourteen-year sentence (reduced to eleven years on appeal) in 2015 following a laborious investigation into the ways in which Libor had been rigged. It's worth noting that he has criticised the authorities, in particular the Financial Conduct Authority and the Serious Fraud Office, for their 'lack of understanding' of how the system worked. He has also issued attacks on the Bank of England and the senior management of the banks for allegedly being complicit in practices such as the lowballing of Libor submissions.

It's impossible to know how widespread the practice was (or is); numerous other individuals have been tried and acquitted, but given the huge amounts of money that were riding on such an easily influenced heuristic, it seems at least probable that others have had the same idea as Hayes.

There have been extensive reforms to the way that Libor is set, and it has been made a criminal offence in the UK to knowingly or deliberately make false or misleading statements when it comes to setting the benchmark. But the moral of the story is probably a wider one: where a huge amount of money is riding on any small element of the global financial system, there are likely to be individuals looking for dishonest ways to manipulate that element.

Bullet Point: Nick Leeson and Tom Hayes were probably just the ones who got caught.

THE ROOTS OF MIS-SELLING

In theory, insurance is a brilliant mathematical idea. A large number of people pay a small amount into a fund. The actuaries work out the price, based on the established level of risks of a particular misfortune happening to each individual, so the insurance company can ensure a profit. Then when someone's house burns down or they suffer from an accident, they receive the money they need to deal with the consequences.

In practice there are a few problems. The first is inherent to the industry: it involves an unusual relationship between buyer and seller. There is no actual product or service being provided, just redistribution of a pot of money. For every one person who receives a benefit from the scheme, many receive nothing. This motivates some individuals to attempt to cheat the system through fraudulent claims, which raises premiums for everyone. This has consequences: the sheer level of claims where a small car accident is reported as having given the driver 'whiplash' (which is relatively easily faked) has led to whiplash claims being severely limited in the UK, and to any genuine whiplash sufferers being treated with scepticism and short-changed when it comes to their medical treatment.

On the other side of the coin, there is an obvious incentive for the companies themselves to deny legitimate claims and to sell insurance at a price that overestimates the actual risk. So you end up with an arms race between the two dishonest tendencies: the customers who want to maximise their returns and the companies who want to minimise them.

Retail companies have absorbed the more devious lessons of the insurance industry by incorporating service contracts with the products they sell: the profit margin on these service contracts is often as much as four times as high as the margin on the actual product. For instance, a washing machine sold with a three-year warranty and service contract may well have been studied by actuaries who concluded that the risks of failure within three years are extremely low.

And this is the ideal kind of policy for any insurance company or financial organisation to sell: one that is highly unlikely to be invoked. In the UK one of the most egregious cases was the mis-selling of personal protection insurance by banks who issued loans or credit cards. In 2005 the Citizen's Advice Bureau issued a 'super-complaint' about PPI. At the time there were about 20 million PPI policies in action. In theory they covered credit payments in the event of illness or unemployment, but there were huge loopholes in the policies and some people would never have been able to claim: for instance, self-employed customers would not have been covered for a loss of income.

It was reported that as much as 20 per cent of the profit of Barclays, one of the major banks, was coming from PPI, with 70 per cent of PPI income being retained. Only 4 per cent of customers ever made claims, and as many as a quarter of those claims were declined.

Eventually the courts declared that many of the PPI policies had been mis-sold, meaning that customers had been given inadequate or incorrect information when they bought them. The banks were ordered to repay significant amounts of money (it ended up costing them more than £13 billion) to customers whose policies had been mis-sold.

So there was at least some justice in the end. But it is worth bearing in mind that insurance policies may be less useful than they appear to be, and to bear in mind that the entire business of insurance is predicated on selling you a possible future amount of money at a price that deliberately overprices that risk in order to ensure a profit for the provider.

My vet once gave me a good piece of advice about pet insurance. Given how overpriced it is and how many loopholes it involves (for instance, the company won't pay out for a cat who isn't fully up to date with its injections, even if the actual injury is nothing to do with the illnesses they combat), it is better to just put a similar amount in a dedicated savings account each month. That way you provide your own insurance without giving a company a profit margin.

Of course, when it comes to insuring your house against fire, that would be dangerous advice unless you have enough money to write off the loss of a house!

Bullet Point: Always, always read the small print in an insurance policy.

WHEN IS COMPENSATION NO COMPENSATION?

A brief morality tale . . . one huge growth area in the UK in particular after the global financial crisis was in payday loans: loans for smallish amounts (up to about £1,000) that were mainly used by low-income individuals, and which attracted high monthly interest rates if they weren't promptly repaid. Many people got into financial trouble after unwisely getting into a spiral of debt while using such loans. (There has been a similar issue in other countries, especially those who don't have statutory maximum interest rates.)

In 2013 it was reported that the rates charged by the companies were as high as 4,000 per cent to 5,000 per cent per annum. Indeed, the highest rate, charged to customers of Wonga, was 5,835 per cent. Wonga's pre-tax profits that year were £62.4 million. At one point, the consumer money expert Martin Lewis calculated that given a compound interest rate of 4,212 per cent, after seven years a £100 loan would cost more than the entire national debt of the USA to repay. Not all companies charged compound interest, and clearly anyone stuck in that situation would get into financial trouble far sooner than seven years later, but the point remains that these are eye-watering rates. One Manchester loan shark was jailed for illegal money lending in 2014 in spite of pointing out that his rates were actually lower than those charged by Wonga. His lawyer argued that: 'Within his community there are large numbers of people who because of their debt history are very limited in terms of the places they can go, other than companies such as Wonga.'

The degree of the problem attracted a lot of media attention and political criticism and eventually the Financial Conduct Authority stepped in to cap interest rates at 0.8 per cent a day along with a cap on the total amount repayable at 100 per cent of the loan. There were also numerous claims made against the companies on the basis that they had mis-sold their loans, claims that were granted in cases where the companies had not clearly explained that the loans should only be used for short-term cash-flow issues, not as a longer-term solution.

The reforms still allowed companies to charge fairly high rates, but combined with the claims made against them, this undermined their business model to the degree that several of them went out of business, including Wonga.

When it went into administration in the UK, there were nearly 400,000 outstanding claims for an average of £1,200 against Wonga, which had started out as a UK-based company, but had shifted significant parts of its business to Switzerland and continued to operate in other jurisdictions. However, because the UK part of the company had gone out of business, the claims were dealt with by the administrators, meaning they could only be applied to the specific part of the business that was still based in the UK.

In January 2020, it was announced that the final settlement for the claims would be just 4.3 per cent of what the aggrieved customers had been awarded. So those average £1,200 compensation claims turned out to be worth just £64 each.

Bullet Point: If it's legal, someone will do it.

Out of Thin Air

In the 1930s Henry Ford supposedly said that it was a good thing most Americans don't understand how banking works because if they did, there would be a revolution within days. The standard view of banking is something like this: people deposit their money in a bank. Banks lend out a proportion of the money they are holding as loans, whether to businesses or to individuals. They are allowed to lend out more than they actually hold in reserve because of the fractional reserve system, but they can borrow from the central bank to make up the shortfall. Central banks create money but have to be careful not to create too much of it, because if governments could just print money to spend it they would do so irresponsibly. This view treats money effectively as a limited resource, or one we have to be extremely careful not to create too much of, which justifies the idea that there simply 'isn't enough money' to fund certain types of government spending.

For a long time, this was the anodyne view of banking that was propagated by central banks such as the Bank of England. However, in 2014, the BoE published a remarkable paper called 'Money Creation in the Modern Economy', which endorsed a far more radical view. In its summary it stated that: 'Rather than banks receiving deposits when households save and then lending them out, bank lending creates deposits ... In normal times, the central bank does not fix the amount of money in circulation, nor is central bank money "multiplied up" into more loans and deposits.'

To fully understand the consequences of this, let's take a brief look at the history of fractional reserve banking. Sometime around the late fourteenth or early fifteenth centuries, goldsmiths, who had been charging fees for storing gold and issuing notes that could be used to reclaim it, realised they could make more money by lending out some of the gold they were storing, in the form of additional notes. Since the notes were being used as currency, they would usually have enough gold on hand to repay any notes that were brought back to them.

This meant they changed role from acting as a storage vault for gold to being more like a bank, which could create money by lending it out. Think about it: if for every gold coin in storage they have issued two notes to the value of that coin, then they have multiplied the amount of money in circulation by two. (That is the multiplier mentioned in the BoE quote above: the creation of money by the bank multiplies the amount of money in existence.) They have basically created money out of thin air. In this respect, money is merely an IOU.

Of course, some goldsmiths and early banks got into trouble. For instance, if a rumour started that they weren't solvent there could be a bank run in which all depositors demanded to be paid at once, leading to a collapse of the bank. As this problem led to increasing numbers of economic crises, central banks were created, starting in Sweden with the Riksbank, in 1668. They were used as lenders of last resort, in an attempt to stabilise the developing banking system. Of course, this didn't always work, but gradually the modern system of fractional reserve banking became orthodox.

What the BoE were admitting in that paper was that the actual process of money creation is generally carried out by the private banks, and that there is no limit to how much they can create, so long as they can find someone to lend it to. So the basic driver of the modern economic system is interest-earning debt. The main role of the central bank is to lend credence to the system, to set the price of borrowing money through the interest rate, and to try to be a cushion against bank runs and economic crises.

Why do we still believe that money is a real thing that is in limited supply? Well, there is some truth in what Henry Ford said: if people realised that the government really doesn't have any limits on what it can spend, that the loans they have to pay interest on in order to buy a house were created out of thin air, and how fragile the entire system is, then they might be far less willing to comply with the system in its current form. So while the central banks may admit in technical documents that our money supply is mostly created by bank lending, the establishment in general will continue to talk about money as a real, limited substance that is created by the central bank.

It may be a fiction, but perhaps it is a useful one.

Bullet Point: There isn't actually any money in your bank account, only an IOU.

CONCLUSION: IS THE FINANCIAL SYSTEM A PONZI SCHEME?

We've already seen how most money is created by banks lending it out. And we've seen how huge fortunes can be made through small distortions in the global financial system. However, it's worth going beyond that and wondering whether the entire system is actually fundamentally flawed.

It's not hard to find people who claim that the global financial system is actually a kind of Ponzi scheme: that the status quo can only be maintained by the constant creation of more and more money (or monetary equivalents). While some of these claims teeter over into conspiracy theory territory, it is true that the degree to which the derivatives market has grown over the last twenty years is unprecedented. In the US, the big banks were effectively banned from gambling on derivatives until the Glass Steagall Act was repealed in 1999 (following extensive lobbying). Since then it has become an ever larger part of big bank activity around the world. The total outstanding derivative contracts, including interest rate swaps, currency swaps, credit default swaps, commodity futures options contracts and stock market futures options contracts in 2001 was about $100 trillion. Now it is well over $700 trillion, which is about eight times as much as annual global GDP.

Some argue that this is not as scary as it sounds: since there are two sides to every derivatives deal, the apparently huge amount of money kind of 'cancels itself out'. But that doesn't mean a huge amount of damage couldn't be done by a partial collapse in any part of that derivatives market; and given that the banks are now seen as too big to fail, the consequences are likely to fall on nation states, central banks and, down at the bottom of the pile, ordinary taxpayers.

Derivatives are more heavily regulated than they were, but the legislation put through after the 2008 crisis has many loopholes: for instance, the US regulators are given information about Goldman Sachs International, a US company, but Goldman Sachs International Bank is separately regulated by the UK authorities; in spite of this, the latter company is a subsidiary of the US company, and problems at the UK branch could have a significant impact on the parent company.

Werner Bijkerk, the former head of research at the International Organization of Securities Commissions, an umbrella group that attempts to oversee derivatives markets, has pointed out that this means no one has a full picture of the global risks: 'It's a global market, so you really have to have a global set of data . . . You can start running "stress tests" and see where the weaknesses are. With this kind of patchwork, you will never be able to see that.'

This is partly due to a bunfight between regulators in different countries, some of whom protect their own banks' right to not be interfered with or overseen by foreign regulators. But the result is that the level of risk is hard to assess. The banks will argue that they have nothing to hide (they would say that, wouldn't they?) and that regulators have access to any information they want. But derivatives trades are not automatically reported so the onus is on the regulators to guess where the problems might be.

In October 2016, the Commodity Futures Trading Commission suggested a law requiring all American bank subsidiaries to report their derivatives exposure along with regulations to make derivatives trading less profitable, but the banking industry lobbied against it and after the election of President Trump, it wasn't authorised.

So the derivatives market continues to be a source of concern for those who worry about how robust the global financial system is.

There are also other reasons to worry that the global economy is heading for danger. The Nobel Prize-winning physicist Steven Chu has argued that the whole system is flawed as it is entirely dependent on growth: 'Increased economic prosperity and all economic models supported by governments and global competitors are based on having more young people, workers, than older people. Two schemes come to mind. One is the pyramid scheme. The other is the Ponzi scheme.'

The point is that young workers pay the costs for healthcare and pensions for older workers. This requires a constant growth in young workers (which explains why the Chinese government has dropped its one-child policy). But the cost of constant economic and population growth is an ever-increasing use of limited resources. And as countries become more prosperous, the birth rate tends to fall, leaving them more and more reliant on immigration to make up the difference. No one has really come up with a way to maintain current economic standards at a steady economic state without population growth.

Chu says: 'The economists know this, but they don't really talk about it in the open, and there's no real discussion in government. Every government says you have to have an increase in population, whether you do it through immigrants or the home population. So, this is a problem.'

So, do the economists and the bankers really know things that we don't know?

Possibly.

Do they have any long-term solutions?

Probably not.

Bullet Point: Don't worry. We're all doomed in the long run.

People Suck

My eighteen-year-old daughter, who has recently survived the 'difficult' teenage years, has gradually developed the habit of responding to the news about any new lie, outrage or scandal by shrugging and saying, 'People suck,' before turning wearily away.

On one level she is completely right. People do suck. We suck at understanding and successfully applying basic mathematical concepts. We suck because we often prefer to stick to what we already believe rather than impartially reviewing any new evidence that is presented. We suck because we are subject to cognitive biases, biases that make it far too easy for liars, psychopaths and lying psychopaths to mislead us.

And let me make it clear I am not, in any way, setting myself above all this. I am a normal person, subject to exactly the same biases, irrational tendencies and motivated rationality that you or the next person is. I suck too.

And writing or reading a book like this can't help but remind you that there are a lot of truly dishonest people in the world. Whether they are politicians, PR directors, sales guys, advertisers, businesspeople, financiers or whatever . . . A lot of people are self-interested, self-deceptive or self-glorifying. And many of them find ways to use numbers to pursue their own interests, to the detriment of other people.

But I don't want to end on that negative note. I don't want to turn wearily away, and I don't want you to either.

Numbers are our friends. They really can be impartial. There are, as I've discussed, many ways to distort, conceal, shape or ignore them. They can be brushed under the carpet. Their significance can be doubted, maligned or dismissed.

But they can still help us to discover the truth. The crucial thing is that, rather than turning wearily away, we continue to use our critical-thinking faculties. We have to bear in mind our own cognitive biases and try to step outside of them, to interrogate ourselves, to double-check whether or not we are falling for the lies of a lying liar.

If you want to know the truth, the key is to work out how the numbers are being abused, to insist that the real, pure numbers are revealed and to trust what the numbers are telling us. We need to make sure we get the numbers that describe a situation from as many different angles as possible. We need to investigate the motives of the person giving us the numbers and we need to find other sources, differently motivated people, different ways of gathering data, to keep prodding and poking as much as possible rather than passively accepting the numbers we are given.

Always remember that sunlight is the best disinfectant. Liars rely on hiding parts of the situation, drawing your attention away from the things they don't want you to think about, and muddling up the situation so that you don't trust what the numbers say.

So let's not despair of human nature. Instead of letting the numbers be used as a smokescreen, let's trust that they can help to bring the sunlight that is needed into the room, and that the truth can be revealed.

Yes, numbers can be used to lie. And they often are used that way. But that isn't their fault. It's the fault of the people using them.

Numbers don't lie.

People do.

Comparable Titles

BAD SCIENCE
Ben Goldacre, 2009
A broader approach to how science gets things wrong and is misinterpreted.

HOW TO LIE WITH STATISTICS
Darrell Huff, 1954
A classic guide to how statistics can be used to mislead, illustrated with cartoons, now somewhat dated.

IRRATIONALITY: THE ENEMY WITHIN
Stuart Sutherland, 1992
A terrific survey of a variety of forms of irrational belief, including sections on the misuse and misunderstanding of statistics. Opinionated, and a broad philosophical take, but still a good book.

PROOFINESS: THE DARK ARTS OF MATHEMATICAL DECEPTION
Charles Seife, 2010
An excellent academic approach to the art of using numbers deceptively.

THE TIGER THAT ISN'T
Michael Blastland and Andrew Dilnot, 2007
The creator and presenter of Radio 4's *More or Less*, with an entertaining, somewhat metaphorical approach to trying to see through the fog of statistics.

Index

Note: page numbers in **bold** refer to diagrams.